Integrated River Basin Governance

Integrated River Basin Governance

Learning from International Experiences

Bruce P. Hooper

Publishing
LONDON • SEATTLE

Published by IWA Publishing, Alliance House, 12 Caxton Street, London SW1H 0QS, UK

Telephone: +44 (0) 20 7654 5500; Fax: +44 (0) 20 7654 5555; Email: publications@iwap.co.uk
Web: **www.iwapublishing.com**

First published 2005

Copy-edited and typeset by TechBooks Ltd, India

British Library Cataloguing in Publication Data
A CIP catalogue record for this book is available from the British Library

Library of Congress Cataloging-in-Publication Data
A catalogue record for this book is available from the Library of Congress

ISBN: 1 84339 088 4

Contents

Preface

This idea of this book was conceived in the early 1990s in the backblocks of the Namoi Valley. I was doing research in integrated floodplain management on the vast riverine floodplains of the northern Murray-Darling Basin, Australia. I had spent every day for two months interviewing farmers and agency staff about floodplain management.

Everyone had their own solution—but very few had the 'big picture' about the immense valley in which they lived and worked. Why was this? There were so many landscape interdependencies.

After a fifteen-year gestation, this book came together in 2004 in Carbondale, near the confluence of the Ohio and Mississippi Basins.

The growth of this book was like a river: fed by tributary inflows from practical research and consultancy projects in India, Australia, United Kingdom, Canada, New Zealand and the United States. I trust it has now reached some semblance of adulthood.

In some respects, I see this book cataloguing the growth and maturity of integrated river basin management. By the end of the twentieth century, it had become apparent that single issue, sectoral natural resource management in many of the world's great river basins had produced suboptimal outcomes for both people and the natural environment. This happened in both developing and developed countries.

There is now growing acceptance that an integrated, adaptive approach to river basin management is needed, one which harnesses the power of people and the institutions we create. This approach, integrated river basin management (IRBM), is the focus of this book.

Why a book on IRBM? It was written to help advance thinking on the decision-making dimensions of river basin management, and to extend the understanding of how integrated water resources management is implemented at the basin scale. It was also written to help practitioners implement IRBM and to assist a growing generation of river basin managers in professional practice. I trust this book is also useful for postgraduate students, providing practical tools and conceptual frameworks.

As well, this book provides the broader water community with a resource on experiences in river basin management.

This book owes its existence to several people who provided practical assistance. I thank Ian Borthwick, Michael Dunn and Alan Peterson (IWA Publishing) for their collegiate and helpful advice; KVGK Rao (Andhra Pradesh) for assistance in the Warangal case study; Sergio Bögeholz (Sydney) for assistance in the evaluation of catchment management organizations; Helen Sartori and Dianne Dredge (Brisbane) for the evaluation and compilation of information systems; staff of the Andhra Pradesh Water Conservation Mission for photographs and maps; Kristen Milton, Murray-Darling Basin Commission (Canberra), for cartography of that basin; Aaron Wolf, Marcia Macomber and Sam Littlefield, Oregon State University, for providing data and maps of international river basins; Rishath Ahamed, Nathan Eidem, Greg Gurbal, Stacy Nicklow and Michael Pease (Carbondale) for copyediting, style formatting, database management and case studies.

I acknowledge the international contributions of Jenny Bellamy, John Burton, John Fargher, Clive Lyall, Graham Marshall, Jim McDonald, Peter Millington, John Pigram (Australia); Tony Dorcey, Bruce Mitchell, Dan Shrubsole (Canada); KVGK Rao, PS Rao (India); Frank van Steenbergen (The Netherlands); Hilary Sunman, Colin Green (United Kingdom); Steve Born, Ben Dziegielewski, Chris Lant, Bob Lee, Rich Margerum, Tony Prato (USA).

I would like to thank the many reviewers who provided valuable comments on the proposal to write this book.

Finally, I thank my wife Kaye and my children for their encouragement and support.

IRBM is both an art and a science. I trust that readers of this book will be better equipped to manage river basins.

Bruce P. Hooper
August 2005

1

The IRBM Paradigm

1.1 PREVAILING CONCERNS

The relationship between the management of a river basin's land and water resources and the quality and quantity of the downstream water resources is apparent. River basin management is widely accepted as a critical task in providing resource use products and the management of natural resources (Burton 1991; Murakami 1991; Newson and Fang 1991; Newson 1992).

The prevailing concern is that the management methods to produce improved river basin governance need to be better known. These methods vary and depend on the local cultural, political, administrative and institutional context. The methods include on ground actions, plans of management, strategic natural resource management policies, use of resource science, engineering and economic analysis of management options, community participation, incentives and many different types of government and community-led initiatives. These are the essential institutional arrangements relevant to a river basin and form the governance dimensions of natural resources management at the basin scale.

In this book, it is contended that the critical factors which preclude effective river basin management are institutional, organisational, economic and socio-cultural. They refer to mechanisms and issues such as the roles and responsibilities of river basin organisations (RBOs), the management of common-pool natural resource 'commons', property rights, water allocations mechanisms, adoption of sustainable land and water management practices, jurisdictions of governments in

 ISBN: 1 84339 088 4.

water and soil resources, limitations and benefits of public involvement in decision-making and others.

This book is written to assist the improvement of river basin management by focusing on the institutional, organisational, economic and social factors involved in decision-making about river basin management. In this first chapter, I discuss how water resources management has undergone a paradigm shift in the latter part of the twentieth century—from a single sector to integrated approaches. I then examine how the 'watershed' (at the local catchment, river valley or river basin scale) has emerged as a locus for implementing integrated water resources management (IWRM). I then finish by addressing some scale issues in river basin management.

1.2 CHANGING PARADIGMS IN WATER RESOURCES MANAGEMENT

1.2.1 Paradigms reflect societal priorities

A definition of a paradigm is:

> The working assumptions, procedures and findings routinely accepted by a group of scholars, which together define a stable pattern of scientific activity; this in turn defines the community which shares in it.
>
> (Gregory 1994)

New paradigms for water resources management characterised the last decade of the 20th century, but originated in the political and economic development priorities of earlier times. Natural resources management emerged as a planning process to allocate natural resources for required human uses of land and water resources (O'Riordan 1971). Natural resources management has been practised in various forms since the earliest periods of human occupation and settlement. Some of the earliest know irrigation systems in the Nile, Yangtze and Indus Valleys are examples of harnessing riverine resources, a rudimentary form of river basin management. However, it has only been during the last century that the catchment has been the focus of management, brought about by an increasing recognition of the importance of water resources as fundamentals for human use and ecosystem functioning (Pigram 1986; Newson 1992; MacKenzie 1996).

Approaches to water resources management in any period reflect the prevailing government policies and societal values of the day. There were rapid paradigm shifts in water resources management thinking in the twentieth century, from the ethic of resource exploitation, to resource conservation and sustainable resource management. Traditional approaches were essentially hydro-centric. They were single sector (water) oriented in which the river basin or groundwater province was viewed as a complex physical system—based on interrelationships between the hydrological and geomorphologic characteristics of the basin and its rivers and streams. This approach, common in the 1930s to 1960s and favoured by water engineers and water economists, viewed the basin as a water resources system whose water resources were to be exploited for economic development. The approach emphasised determining maximum possible yield and developing mechanisms for most effective water allocation among users. It was used for significant water resources development projects, such as the Hoover Dam project in the USA—an era characterised by dam building and irrigation expansion in very large water resources projects. The single sector approach was driven by highly scientific methods and technological innovation, with an overall purpose of maximising available yield from the river basins of local watersheds. More complex approaches evolved promoting multi-objective development of water resources systems including recreation, hydropower, navigation and irrigation development, as evidenced in the work of the Tennessee Valley Authority and the US Army Corps of Engineers in the USA, the Nagarjuna Sagar Dam project in India and the Snowy Mountains Scheme in Australia. The engineer's and the water manager's view of river basin management is water-centric, based on complex interrelationships between the hydrological and geomorphological characteristics of the river basin and its rivers. The approach is at best multi-objective, endeavouring to achieve synchronous management of water resources.

The water resource is regarded as an infinite resource in which many uses can be achieved from a single water resource stock. Reuse is common, and highly engineered solutions to complex water management are built on the premise for maximising yield and allowing no water to 'go to waste' through ocean ('end of pipe') outflows. This approach is typified by the use of highly sophisticated decision support systems, spatial planning of water resources using geographical information systems, top–down hierarchical command and control management systems, large investments in data collection and management, and strong commitment by governments to support water resources development by political will (including direct intervention) and long term funding programmes.

The environmental movement of the 1970s heralded a new era in water management and questioned these approaches. A new focus on ecosystems based on the new science of ecology, questioned the single or multi-objective approach to water resources management, with its strong development emphasis. The reality was that the traditional paradigm ignored the more diverse range of resource use features of river basins, which interact to create the so-called 'wicked' problems of environmental management and sustainable water resources management. The new paradigm recognised river basins as large, complex, integrated ecological systems. The concern was for ecosystem deterioration and negative social impacts caused by water development projects.

1.2.2 Integrated water resources management—use and definitions

An integrated approach to water resources management (IWRM) emerged in the 1980s but had its origins in this new thinking of ecosystems management and transactional planning which goes beyond the single resource and multi-objective approach paradigm. It emerged as a new paradigm for several reasons. First, traditional environmental management had been largely reactive, disjointed and based on narrow or limited purposes. As (Mitchell 1991) suggests,

> Many if not most of our government natural resource management organizations were not designed to deal with ecological problems which are characterized by inter-linkages and inter-relationships.

Second, many environmental problems have been called 'wicked' problems. They arise from interrelationships among biophysical, human and economic systems, and therefore can rarely be treated in isolation. Finally, increasing resource demands have led to conflicts over environmental management. Governments are finding it increasingly difficult to make environmental management decisions without incurring conflict.

The integrated approach is sometimes called, more generically, integrated resources management (IRM) or integrated resource and environmental management (IREM). The approach is the foundation for international and global environmental management initiatives aimed at more sustainable management, as developed at the World Commission on Environment and Development 1987, and the United Nations Conference on Environment and Development Agenda 21, 1992 (Born and Sonzogni 1995). These global initiatives reflect the prevailing concerns of their time: that co-ordination achieves far more than fragmentation of decision-making in resource use, and that water is fundamental to human life and ecosystem functioning.

Nowhere is this more apparent than in the implementation of the World Water Council's World Water Vision 2000, which strongly endorsed an integrated approach to water management. This initiative has various origins, including:

- International Conference on Water (Mar del Plata, Argentina, 1977);
- World Consultation on Drinking Water and Sanitation (New Delhi, India, 1990);
- Dublin Conference on Water and Environment (1992);
- Rio Summit (Chapter 18 of Agenda 21) (1992);
- Ministerial Conference on Drinking Water and Sanitation (Noordwijk, Netherlands, 1994);
- First World Water Forum (Marrakech, Morocco, 1997);
- Ministerial Conference on Water and Sustainable Development (Paris, 1998);
- Sixth Session of the United Nations Commission on Sustainable Development (1998);
- Workshops and publications leading up to the Second World Water Forum (The Hague, 2000);
- Outcomes of the Third World Water Forum, 2003.

As these dates suggest, IWRM emerged at the international level as the new paradigm in the 1990s. It brought with it concerns for ecological health of rivers, floodplains and river basins which supply water to rivers and groundwater resources, and resource use impacts on the functioning of ecosystems within a watershed. Mitchell (1983) claimed that the challenge is how such an approach is to be interpreted. Advocates of an ecosystem approach have interpreted it to be synonymous with a comprehensive approach in which attention is given to all components and linkages in a system. When a comprehensive approach is taken, the probability is very high that the period of time required to complete an analysis or a plan will be very long, resulting in the final plan often being no more than a historical document, because too many events or processes will have changed and made the plan obsolete before it is even completed. Alternatively, the interpretation of an integrated approach involves a more selective or focused perspective. Not all components and connections in a system are considered, but only those which, on the basis of knowledge from all stakeholders (through focus groups or other forums involving people ranging from technical analysts to long-term residents), are judged to be the key drivers of variability in the system. Both a comprehensive and an integrated interpretation are consistent with an ecosystem approach, but the latter leads to a more focused approach and therefore increases the likelihood of a more practical output (Hooper, McDonald and Mitchell 1999).

Much of the conceptual development and experience with integrated approaches relates, not surprisingly, to water and related land resources, with catchments and bioregions being used as the site of implementation. Such efforts include inter-relating the management of water quality and quantity, ground and surface waters, the land–water interface, biologic concerns and the objectives of the user community.

The IWRM approach uses stakeholder participation, cross agency co-ordination and a wide range of innovative tools to improve water management. These tools are now documented (such as in the Global Water Partnership's ToolBox www.gwpforum.org). It is being increasingly used on a watershed/river basin basis (Environmental Protection Agency 1992; Environmental Protection Authority 1993; Mitchell and Hollick 1993; Hooper 1997; Born and Genskow 1999; Bellamy et al. 2001). IWRM, while a response to past narrow and disjointed approaches to natural resources management, aims to overcome the dysfunctional mechanisms between and within the government and communities in the management of water resources. This participatory approach seeks involvement through negotiation and building partnership agreements. It seeks to avoid marginalising resource user groups or agencies. It builds bridges and partnerships to achieve commonly accepted resource management goals.

An integrated approach to natural resources management is intuitively appealing and reinforces an ecological approach to landuse planning (Mitchell 1989). Many agency natural resource managers, academics, agency professionals, industry organisations, community organisations and resource user groups have supported planning and managing water and related land resources on a watershed (catchment, river basin) basis and the approach is now being widely adopted (Burton 1986; Mitchell 1987; Burton 1988; OECD 1989; Downs, Gregory and Brookes 1991; Born and Margerum 1993; Environmental Protection Agency 1993; Mitchell and Hollick 1993; Rogers 1993; Born 1994; Born and Sonzogni 1995; Anonymous 1997; White 1997; Heathcote 1998; Ballweber 1999; Batchelor 1999; Bellamy et al. 1999; Margerum and Born 2000; Gonzalez and Arias 2001; Jonch-Clausen and Fugl 2001; Hooper 2002; CGIAR Challenge Program on Water and Food 2003).

Examples include:

- The World Water Council which endorsed an integrated approach to water resources management in the World Water Vision, 2025, and used an expert panel to promote sustainable river basin management based on integration. The report provides a five-stage framework for IRBM at the river basin scale: assessment of the institutional framework, co-operation strategies, formation of a river basin management authority and management plan, implementation of the management plan, evaluation and compliance monitoring (Anonymous 2000).
- The Global Water Partnership which developed a Framework for Action to implement IWRM and an IWRM ToolBox containing over 50 methods to implement the IWRM approach (See http://www.hrwallingford.co.uk/projects/gwp.fau/toolbox/).
- The International Network of Basin Organisations (INBO), based in France, established to promote IWRM at the level of river basins and facilitate implementation of tools suitable for the integrated management of water resources at this scale.
- The International River*symposium* and River*prize* based in Brisbane, Australia, which rewards demonstrated achievement in river management with an annual prize of A$100,000 and a symposium of latest river management practices, undertaken within an IRBM approach. The 2000 winner was the Grand River Conservation Authority, Canada (http://www.riverfestival.com.au/2001/symposium/riverprize_2000winner.asp),
- The Stockholm Water Symposium and The World Water Prize which demonstrate and showcase advances in IWRM.
- The International, American and Canadian Water Resources Associations who promote and continue to enhance knowledge of IWRM, and international collaboration between, professional experts and practitioners in IRBM, in countries both of the south and north.

There have been many definitions of IWRM or the integrated approach. These include:

> A process of formulating and implementing a course of action involving natural and human resources in an ecosystem, taking into account the social, political, economic and institutional factors operating within the ecosystem in order to achieve specific societal objectives.

(Dixon and Easter 1986 quoted in Born and Sonzogni 1995)

> The co-ordinated control, direction or influence of all human activities in a defined environmental system to achieve and balance the broadest possible range of short- and long-term objectives.

(Cairns 1991 quoted in Born and Sonzogni 1995)

A more comprehensive or inclusive approach that takes into account the scope and scale of environmental and human issues and their interconnections . . . a strategic and interactive process is used to identify the key elements or goals at which to direct attention. These critical elements or goals then become the focus of an inter-organizational or coordinated approach to reforming environmental decision-making.

(Queensland Department of Natural Resources 1991)

Proactive or preventative measures that maintain the environment in good condition for a variety of long-term sustainable uses. Alternatively . . . co-ordinated control, direction, or influence of all human activities in a defined environmental system . . . to achieve and balance the broadest possible range of short- and long-term objectives.

(Scherer 1994 quoted in Born and Sonzogni 1995)

A process which promotes the co-ordinated development and management of water, land and related resources, in order to maximise the resultant economic and social welfare in an equitable manner without compromising the sustainability of vital ecosystems.

(Global Water Partnership Technical Advisory Committee 2000)

The similar theme running through these definitions is *co-ordination*.

1.2.3 Integrated water resources management—a cross-sectoral, coordinated approach

IWRM uses co-management but is fraught with classic problems of commonly managed resources: differing interpretations of property rights, conflicts over use, spatial and temporal variations in access to water, susceptibility to hazards of water surpluses or deficits, lack of ongoing financing when other spending (military, health, education) consume public service delivery budgets and others. Despite these problems, IWRM provides mechanisms for meeting top–down with bottom–up management. 'Entry points' for success in IWRM need to be crafted in any geographical setting: primarily though improved human and organisational capacity, dedicated and sustained funding using cost-sharing, water visioning, not just ownership of the 'commons' problem but also covenants of mutual responsibility and self responsibility, and building leadership skills. These are discussed further in Chapter 7.

In practice, IWRM must bring together a diverse array of people who have a 'stake' in a system to collaboratively manage the activities and impacts. These stakeholders include government entities, community organisations, business and industry organisations and other organisations, and individuals with a particular concern or interest in water resources management. IWRM must also involve 'the public' who also have a stake, albeit less well defined. This participatory approach can help produce strategies that are more coordinated, more cognizant of interconnections and more inclusive of the diversity of goals. Furthermore, it is suggested that a collaborative approach produces greater support and commitment, and increases the likelihood of implementation.

Burton (1993) refers to integrated approaches as being about co-ordination and co-operation, not amalgamation. It is about taking a holistic view, and managing specific resource management problems in that context. Quoting Mitchell and Hollick (1993), Burton recognised three dimensions to integrated approaches: a philosophy, a process and a product. The *philosophy* involves fostering an organisational culture and associated attitudes that view collaboration and co-operation as essential. The *process* should be well understood by all players in a catchment management setting. The *products* of catchment management are numerous and vary, ranging from a new process of development of guidelines, to policy products, to a catchment management plan. The critical 'P' of the three is Process. The focus of this book is improving process.

The conceptual development of IWRM was extended recently by the Global Water Partnership (Global Water Partnership Technical Advisory Committee 2000; Jonch-Clausen and Fugl 2001) and international endorsement of IWRM is now seen at the highest levels, such as the 2003 Summit on Sustainable Development, Johannesburg and the 2nd (2000) and 3rd (2003) World Water Forums. In February 2003, at the 3rd World Water Forum in Kyoto, Japan, a Statement, 'IWRM and the Basin Management Theme' was issued. This statement recognised, inter alia, that

> the key issue confronting most countries today is that of effective governance, improved capacity and adequate financing to address the increasing challenge of satisfying human and environmental requirements for water. We face a governance crisis, rather than a water crisis. Water governance is about putting IWRM with river and lake basin management and public participation as critically important elements, into practice.

The statement calls for action in

> new policies, strategies and laws for water resources development and management . . . in a large number of countries, using the principles of IWRM. Such plans have often led to restructuring of the institutional framework as a result, including river and lake basin organizations as the basic institutional entities for implementing IWRM.

1.3 INTEGRATED RIVER BASIN MANAGEMENT

1.3.1 A definition

The *watershed/basin approach* has been equated to *the application of IWRM at regional scales,* focusing on the critical needs of available quantities of clean water for human survival and sustainable development (Global Water Partnership Technical Advisory Committee 2000). Such efforts include inter-relating the management of water quality and quantity, with ground and surface waters, the land–water interface, ecological concerns, economic development and water-related human health. In many countries of both the north and south, IRBM advocates addressing gender issues, and the individual's right to clean, accessible and affordable water from a river basin's resource stock. It supports the use of the best management practice concept, one in which best management practices (BMPs) are developed using affordability, accessibility, appropriateness and equity criteria. Poverty issues continue to preclude effective water management and health improvements, and must be considered in integrated water resources management at the basin scale.

IRBM is a subset of IWRM. It is how IWRM is worked out nationally or internationally across borders at the river basin scale. IRBM is defined as *an integrated and coordinated approach to the planning and management of natural resources of a river basin, one that encourages stakeholders to consider a wide array of social and environmental interconnections, in a catchment/watershed context.* It is different from traditional multi-purpose resource management as it addresses a broader set of issues including social impacts, varying social values and ecosystem functioning. IRBM implies the inclusion of a full array of physical, biological and socioeconomic variables involved in managing a hydrologic region for environmental values and human use.

In practice, IRBM brings together a diverse array of people who have a 'stake' (a bargaining position) in a river basin in a process to collaboratively manage the activities and impacts or resource use. These stakeholders include government entities, community organisations, business and industry organisations and other organisations, and individuals with a particular concern or interest. IRBM also involves the general public who also have a stake, albeit less well defined. This participatory approach can help produce strategies that are more holistic, more cognizant of interconnections and more inclusive of the diversity of goals. Furthermore, it is suggested that a collaborative approach produces greater support and commitment, and increases the likelihood of implementation.

There are many terms used: integrated catchment management, integrated river basin development and management, ecosystem management, integrated watershed management, ecosystem management, integrated resource and environmental management, the Watershed Approach and Total Catchment Management (capitalisation deliberate—referring to some countries' terminology).

1.3.2 Other IRBM definitions

There have been many definitions of integrated river basin (or catchment, or watershed) management. These include:

> Total Catchment Management involves the co-ordinated use and management of land, water, vegetation and other natural resources on a catchment basis. ... the government seeks to balance resource utilisation and resource conservation through the minimisation of land and soil degradation and the maintenance of water yield and quality.

(New South Wales Soil Conservation Service 1986)

> Operating at 3 different scales: the bioregion, the watershed, and the watershed sub-basin or subwatershed, the Watershed Approach uses the watershed defined hydraulically and hydrologically as the primary boundary for an ecosystem approach to landuse planning. Where possible, the impact of landuse changes or proposed developments will be evaluated on the basis of their

impacts on the watershed, subwatershed, and aquifer system, including upstream/downstream and cumulative effects of these changes.

(Ontario Ministry of Environment and Energy 1993)

The aim of the Watershed Protection Approach is:

To meet water quality goals by using a watershed-oriented approach . . . a comprehensive approach that takes into account all threats to human health and ecological integrity within specific watersheds . . . and involves problem identification, stakeholder involvement, integrated actions, measuring success.

(Environmental Protection Agency 1992)

The co-ordinated management of land and water resources within a river basin, with the objectives of controlling and/or conserving the water resource, ensuring biodiversity, minimising land degradation, and achieving specified and agreed land and water management, and social objectives.

(Hooper 1997)

River basin development and planning (RBDPM) is much more than water resources development, seeking to integrate three interrelated, but separately evolved concepts: (a) multi-purpose development; (b) an integrative role for the drainage basin unit; (c) the acceptance of intervention to promote development (typically seen to be improvement of social welfare or regional conditions).

(Barrow 1998)

Murray–Darling Basin Commission Goal Statement:

Integrated catchment management (ICM) is a process through which people can develop a vision, agree on shared valued and behaviours, make informed decisions and act together to manage the natural resources of their catchment. Their decisions on the use of land, water and other environmental resources are made by considering the effect of that use on all those resources and on all people within the catchment.

The decision to manage our natural resources on the basis of catchments reflects the importance of water to the Basin environment, and to the people who live and work within the Basin.

The boundaries for ICM in the Basin are based on catchments, but in some cases also take account of political, economic and social boundaries.

(Murray–Darling Basin Commission 2001)

These definitions reveal a common thread: that planning the management of water resources is best done using coordinated planning and decision-making across different jurisdictions, with the catchment being the management unit.

A river basin is best viewed as an integrated ecological system which produces natural resources, products or amenities of direct or indirect human value and ecosystem services of intrinsic worth (Burton 1991). Integrated land and water management aims to manage the system in an integrated and holistic way with the objective of maintaining its overall resource productivity on a long-tern, sustained-yield basis. As for IWRM, the IRBM concept provides a way in which an 'ecological' approach can be developed and solutions to complex natural resource management problems are developed in ways that go beyond traditional multi-purpose planning approaches. IRBM attempts to address economic development and resource use from an ecological perspective and identify the sustainable limits of resource use. These are identified as decline in the condition of ecological indicators, beyond which ecological conditions change, perhaps irrevocably, and therefore are nonselfsustaining. IRBM also addresses fundamental human requirements such as assuring accessible, affordable, equitable and appropriate uses of water. This is not an 'add on' but, as for ecosystem health, an integral dimension of basin-level management.

The need for a more holistic approach to managing basin-wide natural resources has arisen because of a poor understanding of the conceptual framework for effective integration, and weak institutional arrangements for integration (Mitchell 1989; Born and Margerum 1993). The demand for a new river basin management paradigm has been driven by ineffectual or unsatisfactory, often undesired outcomes. The lack of success results from incremental and vertically, horizontally, and functionally fragmented efforts to address complex, i.e., 'wicked' problems—problems characterised by substantial scientific uncertainty (Born 1994).

Much of the conceptual development and experience with IRBM relates, not surprisingly, to water and related land resources management with catchments considered generally equivalent to production units (Burton 1986; Naiman 1992). Whether at the local watershed or the very large river basin (such as the Nile), an integrated approach is seen as being most effective as a river catchment represents a classic example of a naturally defined natural resources ecosystem. The catchment boundary clearly defines the hydrological and geomorphological history of ecosystems within the boundary, and establishes a natural barrier for the many biophysical and natural processes at work within it (Burton 1991). The understanding of these processes forms the fundamental principles upon which natural resources management can be achieved.

1.3.3 IRBM elements and characteristics

IRBM involves the following critical elements and characteristics. Mitchell and Hollick define five building blocks for IRBM (Box 1.1). These have been partially modified and can be used to assist process development by setting parameters for the design of management instruments. Similarly, a review of the Wisconsin experience in integrated natural resources management can be used to assist process development, where Margerum (1995) identified a number of elements: IRBM was not a product or an output; it was not just a management technique or strategy; it must include many people and perspectives; it is not a 'quick fix' but takes time; and it requires a long-term view. A third perspective of critical elements was developed by Naiman (1992). He recognised six essential conditions to achieve IRBM (Box 1.2).

Box 1.1 Building blocks of IRBM

1. *Use of a Systems Approach* in which attention is directed towards both natural and human systems, their component parts, and the interrelationships among those parts.
2. *Use of a Strategic Approach* in which attention is directed to key, not all, issues and variables identified through consultation with stakeholders and to linkages among the key issues and variables.
3. *Use of a Stakeholder Approach* in which it is recognised that citizens and non-government groups should be able to participate in decisions about resource management.
4. *Use of a Partnership Approach* in which state governments, local governments and non-government organisations and individuals each have a role, requiring common objective setting, definition of roles and responsibilities, and conflict resolution mechanisms.
5. *Use of a Balanced Approach* in which concerns for economic development are weighed against ecosystem protection, and satisfying social norms and values.

Source: Adapted from Mitchell and Hollick (1993).

Box 1.2 Essential conditions for implementing new approaches to watershed management

- The scope of issues demands unparalleled cooperation between industry, governmental agencies, private institutions and academic organisations.
- The increasing tendency to resort to technical solutions (e.g. hatcheries, silviculture) must be augmented with increased habitat protection and preservation of fundamental components of long-term watershed vitality.
- The complexity of information management and the scope of experimental manipulations needed often exceed the capacity of individual institutions.
- The current tendency to seek conceptual solutions at the expense of data-driven decision must be reversed.
- Intra- and inter-agency inconsistencies in environmental regulations must be corrected.
- Human activities are a key element of ecosystem vitality and must be integrated with environmental considerations before long-term sustainability of the biosphere can be achieved.

Source: Naiman (1992).

IRBM is then characterised as:

(1) *Coordinated* activities rather than amalgamated programmes of action;
(2) Top–down management meeting bottom–up management;
(3) *Strategic planning* rather than all-embracing efforts: being *targeted* and *selective* about actions and prioritising work programs;
(4) *Integrating goals* rather than planning resource use and conservation from either single or multi-purpose reasons;
(5) *Proactive* rather than reactive resource use *planning*: looking to identify problems before they occur and being cautious in resource use;

(6) Using *cost-effective* rather than prescriptive financial management mechanisms;
(7) Using *partnerships* and *cost-sharing programmes* wherever possible;
(8) Working with partners in a co-operative work environment, rather than using confrontational and directive management;
(9) Encouraging *commitment* in staff rather than using command-and-control management styles;
(10) *Empowering* local and regional decision-making rather than centralising decisions and directing staff;
(11) Management based on *problem-solving* rather than functionality;
(12) Having *flexible organisations* rather than rigid inflexible structures;
(13) Providing *appropriate, relevant, affordable* information that is relevant to IRBM;
(14) Using *equitable management methods* which are sensitive to and respect cultural needs and gender issues, and do not discriminate against catchment managers because of their distant location from the decision-making processes of other water professionals.

The practical experience of operationalising IRBM reveals valuable learnings and these are discussed in Chapter 2. However, an overall perspective on how IRBM can be achieved is seen in the following example. An expert group on river basin management prepared a new paradigm for Integrated River Basin Management for the Second World Water Forum and Ministerial Conference of the World Water Council's World Water Vision 2000 in Hague. They maintained that water is an environmental resource, not just a resource for economic production, and water access and security is the basis for social and economic development. River basins are the paramount source of freshwater, and to preserve and maintain this precious resource for present and future generations there is the need for sustainable river basin management. The working group maintained that *political leadership and commitment*are crucial for sustainable water use in river basins. As there are many regional differences in water supply, availability, cultural values, gender issues, technological levels and finance availability, a blueprint for IRBM cannot be given. However, the expert group did advocate the following five elements (Box 1.3).

These elements and characteristics suggest that an approach for IRBM implementation should obviously be *process* driven, one with which clear strategies for action are developed. IRBM should be used as a set of techniques to co-ordinate land and water resources management in a catchment. It should recognise that it is more than applying science to management, but be dependent on rigorous science, resource engineering, rural and urban sociology and resource economics to define and refine good river basin management practices. 'Good' IRBM will depend on 'good' biophysical science, that is, knowing the biophysical characteristics of the catchment and how they operate. 'Good' IRBM will depend on a rigorous understanding of the institutions, the economic forces operating in a catchment

Box 1.3 Elements of integrated river basin management

1. *Basin-wide planning.* Basin-wide planning should balance all user needs for water resources, in the present and the long-term, and should incorporate spatial developments. Vital human and ecosystem needs have to be given special attention.
2. *Participation in decision-making.* Local empowerment and public and stakeholder participation in decision-making will strengthen river basin management.
3. Demand management. Demand management has to be part of sustainable water management. Managing the demand for water rather than continual expansion of water supplies will more likely achieve sustainable use.
4. *Compliance.* Compliance monitoring and assessment of commitments under river basin agreements and arrangements need to be developed.
5. *Human and financial capacities.* Long-term development of sufficient human and financial capacity is a necessity.

Source: Anonymous (1999).

with resulting landuse practices, and the attitudes, behaviours and sociological characteristics of resource users.

There have been a limited number of attempts to build a conceptual framework for IRBM. However, one which captures the essence of an IRBM approach was the innovative framework developed in Australia by Syme (1995)—called PRIME—The Procedural Framework for Catchment Management (Box 1.4).

1.4 SCALE ISSUES—RIVER BASINS AS MANAGEMENT UNITS

There is commonly no one entity responsible for the majority of decisions in river basin management, although in the past and today, organisations such as river basin authorities and commissions have attempted this co-ordination. The scale of application of IRBM decisions is frequently broad, being at the regional and river basin scale. Here strategic decisions are made by governments to allocate natural resources (the macro and meso levels), rather than small-scale (micro) subcatchment approaches. Examples of strategic, regional decision-making include water resources planning for the allocation of urban and rural water (irrigation) from river systems, allocation of mineral resources for extraction from mineral resource provinces and planning timber production volumes and conservation reserves in bioregions.

IRBM involves taking the ‘big picture’ rather than focusing on site specific management issues (Burton 1993), with the latter being developed in the context of and in congruence with the bigger picture of natural resources management in a river valley. The concern is to focus not on the minutiae of specific local actions, which are part of subcatchment management plans. Figure 1.1 shows a Canadian example of the hierarchy of allocation decisions in water resources management.

Box 1.4 PRIME—The Procedural Framework for Catchment Management

Planning
1. Define the problem/scope the issues
2. Collate available knowledge
3. Identify community objectives
4. Identify State Government objectives
5. Negotiate specific objectives
6. Identify issues/knowledge gaps/implementation issues
7. Devise basic catchment plan
8. Identify resources
9. Develop more detailed plan
10. Develop specific evaluation criteria and monitoring indices

Research
1. Identify feasible solutions
2. Identify barriers to adoption of solutions
3. Conduct collaborative action research
4. Identify basic physical and social research needs
5. Conduct basic research

Implementation
1. Derive an implementation strategy
2. Assign priorities & responsibilities for implementation
3. Define available resources
4. Allocate resources for priority activities
5. Conduct and coordinate implementation
6. Design monitoring & evaluation
7. Assign responsibilities and resources
8. Conduct monitoring & evaluation

Source: (Syme 1995).

The focus of IRBM discussed in this book is the Regional and Planning Unit scales shown in this diagram.

1.5 CASE STUDY: LEARNING FROM THE PAST—INVENTING THE FUTURE IN AUSTRALIAN RIVER BASIN MANAGEMENT

Catchments have been the focus of natural resources management in Australia since water shortages were experienced in Sydney in the 1790s. Catchments emerged as a significant focus of land and water management in the 1930s to protect urban water supplies. Recent catchment management emphasises the integrated management of land and water resources over a watershed area for multiple purposes. The scale of application has varied: from small-scale subwatershed efforts, to subcatchment watershed management activities (government protected water

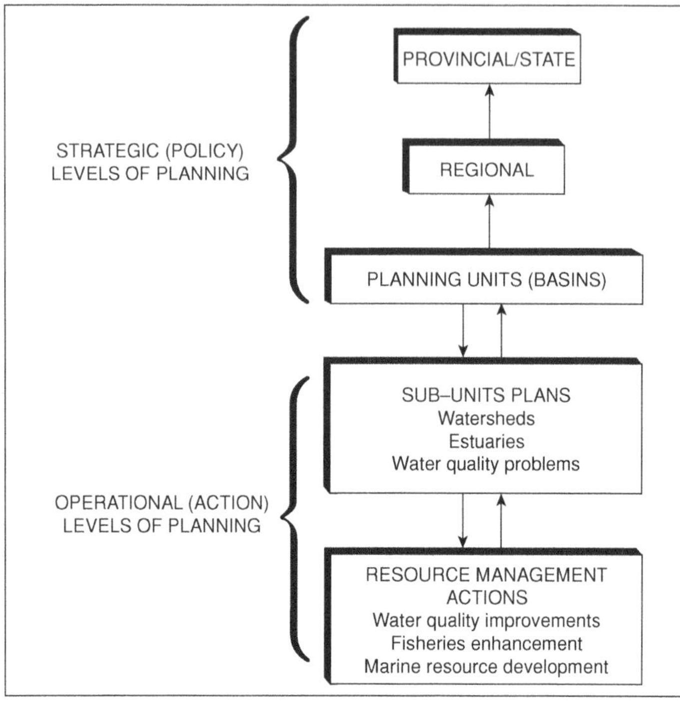

Figure 1.1 O'Riordan's scaled structure of an idealised Canadian water management scheme (adapted from O'Riordan (1986)).

catchments for potable water supply), to river valley catchment management strategies (emerging in some States) and to large-scale river basin management (such as the Murray–Darling Basin Commission).

IRBM is undergoing development and refinement in Australia at the national and regional levels, and has the potential to relate engineering, environmental, social and legal considerations into a more effective system for the management of land and water resources. IRBM approaches are being more widely used throughout Australia. In recent years, substantial gains have been made in Victorian Landcare programmes, the original New South Wales Total Catchment Management programme and more recently the development of an Integrated Catchment Management approach to resource management in Queensland and other states. At the national level, the Murray–Darling Basin Commission initiated an integrated approach in its 1989 Natural Resources Management Strategy and the 2001 Murray–Darling Basin Initiative. Despite the general endorsement of catchment-based management in Australia, there is considerable variation in the procedures and structures in place.

Much can be learnt from the practical experience of turning a concept into reality. The Australian experience perhaps captures some of this experience clearly.

It demonstrates very large areas of natural resources management and extensive, ongoing stakeholder commitment and participation over at least twenty years.

1.5.1 The Murray–Darling Basin experience

An overall appreciation of the relevance and applicability of the IRBM approach can be gained from features of catchment-based management in the Murray–Darling Basin. Said to be 'the largest integrated catchment management program in the world', the Murray–Darling Basin covers an area of over one million square kilometres (Figure 1.2).

The region comprises Australia's largest and most developed river system, covering parts of four States and the Australian Capital Territory in the southeast of the continent. Responsibility for equitable, efficient and sustainable management of water, land and other environmental resources in the basin rests with the

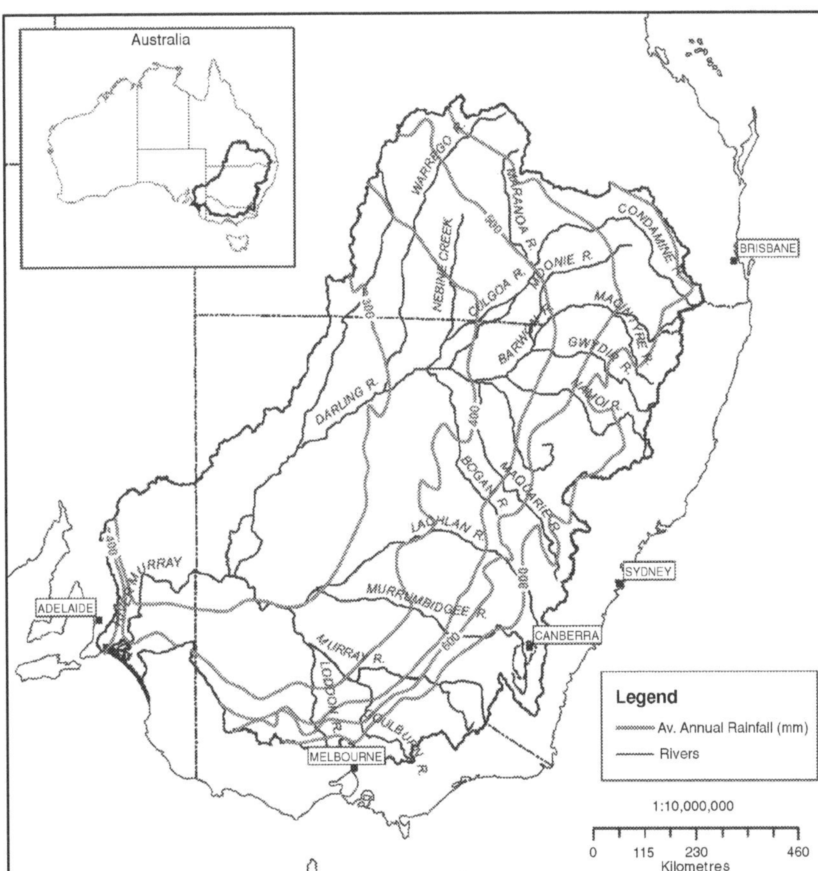

Figure 1.2 Location of the Murray–Darling Basin, Australia.

Murray–Darling Basin Ministerial Council and its executive arm, the Murray–Darling Basin Commission, established in 1986.

Over the past decade, the governments of the region have worked together to achieve substantial progress in co-ordinating actions to deliver positive change within individual catchments and the Basin itself. Despite this progress, less has been achieved than was hoped for. Pressures continue to cause conflict both within and outside the Basin, arising from:

> Increasing competition between agricultural, urban and environmental sectors for the scarce water resources of the Basin; and continuing decline in water quality and ecosystem health as a consequence of past and continuing mismanagement of the Basin's resource base.
> (Murray–Darling Basin Ministerial Council 2001)

In order to reverse the perceived declining trend in the health of the Basin's rivers and ecosystems, The Murray–Darling Basin Ministerial Council and Commission developed an approach to integrated catchment management based on establishing targets for water quality, water sharing, riverine ecosystem health and terrestrial biodiversity, at basin, catchment, sub-catchment and property levels. The new approach is set out in a policy document – 'Integrated Catchment Management in the Murray–Darling Basin 2001–2010: Delivering a Sustainable Future'. The approach replaced the earlier Natural Resources Management Strategy and its implementation features.

- Strengthening institutional arrangements for decision making, especially at the catchment level.
- Building the capacities of catchment organisations to meet agreed targets.
- Identifying the most effective mix of mechanisms to achieve targets and improve Basin health.
- Improving partnerships between governments, communities, and industries.
- Integrating natural resources management with regional strategies and local action plans.

(Murray–Darling Basin Ministerial Council 2001)

Monitoring, evaluating and reporting are essential elements of the Integrated Catchment Management Policy to determine progress in meeting agreed targets for the condition of the Basin's natural resources and the outcomes of investment programmes. The Commission believes that the new policy will deliver improved Basin health and enable all stakeholders to contribute in meeting management targets in a measurable and accountable manner.

1.5.2 The 2000 National Inquiry into Australian catchment management

The Murray–Darling Basin is unique, both in the resource pressures it experiences and in the management response to those pressures. Elsewhere in Australia, there are marked differences in the approach to IRBM, related to the characteristics of individual basins, catchments and catchment communities. In order to develop a consistent approach to ecologically sustainable use of Australia's catchment systems, the Federal Government established in 2000 an Inquiry to report on how best to achieve a nationally coordinated catchment management programme. Specifically, the Inquiry looked at:

- The way in which catchment-based management has evolved in Australia and its value to environmental sustainability;
- Best practice methods for preventing and reversing environmental degradation in catchments and achieving improved environmental health;
- The contribution of different levels of government, the private sector and the community in the management of catchments;
- Options for planning, researching, implementation, co-ordination and cooperation in catchment management;
- Mechanisms for monitoring, evaluating and reporting on catchment programmes.

(House of Representatives Standing Committee on Environment and Heritage 2000)

Some obvious canvassing of the same ground as the Murray–Darling Commission policy was to be expected. However, the Inquiry raised some new issues and offered some fresh initiatives. The report of the Inquiry recommended that the Commonwealth (Federal) Government adopt the lead role in implementing catchment management in Australia. This would entail an agreement on appropriate legislative and institutional arrangements to coordinate the ecologically sustainable use of catchment-based systems at a national level. Working through the Council of Australian Governments, a National Catchment Management Authority would be established to facilitate co-ordination and consistency across a broad range of initiatives. These include:

- Options for a national body of law,
- Accreditation of catchment management plans,
- Education programmes and exchange of information,
- Funding of research.

Provision was made in the report for consultation with stakeholders and for monitoring of compliance with nationally mandated principles, and with priorities and programmes by specified and mandatory target dates, commencing in 2002. An interesting recommendation suggested that, as and when local government boundaries are revised, they be aligned, as far as practicable, with natural divisions within catchments.

The Inquiry recognised the need for funding support to ensure that the problems facing Australia's catchment systems are addressed in a coordinated manner. Several funding options were examined including increases in taxation incentives and removal of disincentives to foster ecologically sustainable use of resources by private sector interests. The preferred option was an environmental levy to pay for the public contribution towards environmental restoration and management. Taxation levies are a popular means whereby governments in Australia have raised revenue for particular purposes.

The Inquiry estimated that a very modest levy, grading from 0.75% to 1.50% of taxable income would raise almost $A4000 million annually. Over 25 years, this should generate sufficient funds to correct environmental degradation and implement sustainable catchment management. However, both the levy and a national catchment management authority were not implemented.

1.6 SUMMARY

In this chapter, we examined how the move from single issue resource management, to multi-purpose and now integrated approaches to water resources management have emerged. IRBM is seen as a subset of integrated water resources management.

There are differing approaches to river basin management throughout the world today, but the preferred approach is to visualise the river basin as an integrated ecological system, which transforms natural inputs (resource stocks), and other environmental components, along with human-made inputs of labour, capital, material and energy into river basin outputs (as water, food, fibre, timber, building materials, minerals etc.). This is a mechanistic view of river basins and their management. Others see river basin management as a type of communicative, transactional adaptive management: a form of dialectical interaction between competing jurisdictions and entities, in which new policy options are tried, lessons learned, and management is refined and improved. Here river basin management is an adaptive, consensual, collaboration building exercise, brokering deals for plans of action. While there are differing approaches to river basin management, there is acceptance of a common philosophy: the use of an integrated approach to land and water management on a watershed basis is essential. The development of this adaptive, integrated approach is discussed further in Chapter 3.

To summarise, IRBM involves improving the governance of river basins. This is done by a range of mechanisms (Chapters 4–7) but initially we will examine the types of river basin organisations and the evaluations of river basin and catchment management to learn from these experiences and find suggested ways to improve governance.

1.7 REFERENCES

Anonymous. 1997, Watershed Management is Key to Improving America's Water Resources. Washington, DC, USA, National Academy Press.

Anonymous. 1999, River Basin Management and Planning: Keynote Paper for International Workshop on River Basin Management. Delft, The Netherlands, RBA TU Delft.

Anonymous. 2000, Towards Water Security: A Framework for Action. Stockholm and London, Global Water Partnership.

Ballweber, J. A. 1999, A critique of watershed management efforts in the Lower Mississippi alluvial plain. Journal of the American Water Resources Association 35[3], 643–654.

Barrow, C. J. 1998, River basin development planning and management: A critical review. World Development 26[1], 171–186.

Batchelor, C. 1999, Improving water use efficiency as part of integrated catchment management. Agricultural Water Management 40[2–3], 249–263.

Bellamy, J. A., McDonald, G. T., Syme, G. J., & Butterworth, J. E. 1999, Evaluating Integrated Resource Management. Society and Natural Resources 12, 337–353.

Bellamy, J. A., Walker, D. H., McDonald, G. T., & Syme, G. J. 2001, A systems approach to the evaluation of natural resource management initiatives. Journal of Environmental Management 63[4], 407–423.

Born, S. M. 1994, Towards Integrated Environmental Management: A Reconnaissance of State Statutes. Boulder, Colorado, University of Colorado Law School. Occasional Paper Series.

Born, S. M. & Genskow, K. D. 1999, Exploring the "Watershed Approach"-Critical Dimensions of State–Local Partnerships: The Four Corners Watershed Innovators Initiative Final Report. Madison, WI, USA, Department of Urban and Regional Planning, University of Wisconsin-Madison.

Born, S. M. & Margerum, R. 1993, Integrated Environmental Management: Improving the Practice in Wisconsin. Madison, WI, USA, The Department of Urban and Regional Planning at the University of Wisconsin-Madison.

Born, S. M. & Sonzogni, W. C. 1995, Integrated environmental management: Strengthening the conceptualization. Environmental Management 19[2], 167–181.

Burton, J. R. 1986, The total catchment concept and its application in New South Wales. Hydrology and water resources Symposium. Griffith University, Brisbane.

Burton, J. R. 1991, Integrated catchment management and its links with public works facilities. Unpublished paper presented to the Asia-Pacific Public Works Conference.

Burton, J. R. 1993, The Big Picture. Water October, 3.

Burton, J. R. 1988, Catchment management in Australia. Civil Engineering Transactions 30[4], 145–152.

CGIAR Challenge Program on Water and Food. 2003, Integrated Water Management Systems.

Downs, P. W., Gregory, K. J., & Brookes, A. 1991, How integrated is river basin management? Environmental Management 15[33], 299–309.

Environmental Protection Agency. 1992, The Watershed Protection Approach. Annual Report. Washington, DC, US Environmental Protection Agency.

Environmental Protection Agency. 1993, Watershed Protection: Catalog of Federal Programs. Washington, DC, USA, EPA.

Environmental Protection Authority. 1993, Geographic Targeting: Selected State Examples. Washington, DC, Environmental Protection Authority, USA.

Global Water Partnership Technical Advisory Committee. 2000, Integrated Water Resources Management. TEC Paper #4. Stockholm, Global Water Partnership.

Gonzalez, A. C. & Arias, C. 2001, The incorporation of integrated management in European water policy. IAHS Publication [272], 69–74.

Gregory, D. 1994, 'Paradigm' (definition). Johnston, R. J., Gregory, D., & Smith, D. M. (eds.). The Dictionary of Human Geography. 3rd Edition. Oxford, Blackwell Publishers.

Heathcote, I. 1998, Integrated Watershed Management: Principles and Practice. New York, Wiley.

Hooper, B. P. 1997, Improving watershed management using an innovative integrated resources management approach in Australia. Journal of Professional Issues in Engineering Education and Practice 123[2], 57–61.

Hooper, B. P. 2002, Towards more effective integrated watershed management in Australia: Results of a national survey, and implications for urban catchment management. Water Resources Update 100, 28–35.

Hooper, B. P., McDonald, G. T., & Mitchell, B. 1999, Facilitating integrated resource and environmental management: Australian and Canadian perspectives. Journal of Environmental Planning and Management 42[5], 746–766.
House of Representatives Standing Committee on Environment and Heritage. 2000, Co-ordinating Catchment Management: Report of the Inquiry into Catchment Management. Canberra, Australia, Commonwealth of Australia.
Jonch-Clausen, T. & Fugl, J. 2001, Firming up the conceptual basis of integrated water resources management. International Journal of Water Resources Development 17[4], 501–510.
MacKenzie, S. H. 1996, Integrated Resource Planning and Management. Washington, DC, USA, Island Press.
Margerum, R. D. 1995, Integrated management for the Wisconsin Department of Natural Resources: An overview. Wisconsin, Wisconsin Department of Natural Resources. Bureau of Water Resources.
Margerum, R. D. & Born, S. M. 2000, A Coordination diagnostic for improving integrated environmental management. Journal of Environmental Planning and Management 43[1], 5–21.
Mitchell, B. 1983, River basin planning: Problems, opportunities and implications for Ontario conservation authorities. Canadian Water Resources Journal 8[1], 1–20.
Mitchell, B. 1987, A Comprehensive-Integrated Approach for Water and Land Management. [No. 1]. Armidale, NSW. Australia, Centre for Water Policy Research, University of New England. Occasional Paper.
Mitchell, B. 1989, Geography and Resources Analysis. 2nd Edition. New York, Longman.
Mitchell, B. 1991, Resource Management and Development. Oxford, UK, Oxford University Press.
Mitchell, B. & Hollick, M. 1993, Integrated catchment management in western Australia—transition from concept to implementation. Environmental Management 17[6], 737–743.
Murakami, K. 1991, River Basin Water Quality Management in Japan—An Overview. Water Science and Technology WSTED4, Vol. 23 [1/3, p 29–39. 1 tab, append.].
Murray–Darling Basin Commission. 2001, Integrated Catchment Management in the Murray–Darling Basin. 2001–2010. Delivering a Sustainable Future. Canberra, Murray–Darling Basin Ministerial Council.
Murray–Darling Basin Ministerial Council. 2001, Integrated Catchment Management in the Murray–Darling Basin. 2001–2010. Delivering a Sustainable Future. Murray–Darling Basin Ministerial Council.
Naiman, R. J. 1992, Watershed Management: Balancing Sustainability and Environmental Change. New York, Springer-Verlag.
Newson, M. 1992, Land, Water and Development. River Basin Systems and their Sustainable Management. New York, USA, Routledge.
Newson, M. & Fang, Z. 1991, Catchment control and planning: emerging patterns of definition, policy and legislation in UK water management. Land Use Policy 8[1], 9–15.
New South Wales Soil Conservation Service. 1986, Total Catchment Management. A State Policy. Sydney, NSW Government.
O'Riordan, T. 1971, Perspectives on Resource Management. London, Pion.
J. O'Riordan. Some examples of land and water use planning in British Columbia, Canada. In: *Land and Its Uses—Actual and Potential*, edited by F. T. Last, M. C. B. Hotz, and B. G. Bell, New York: Plenum, 1986, p. 193–211.
OECD. 1989, Water Resource Management: Integrated Policies. Paris, OECD.
Ontario Ministry of Environment and Energy. 1993, Subwatershed Planning. Toronto, Queen's Printer of Ontario.
Pigram, J. J. 1986, Issues in the Management of Australia's Water Resources. Melbourne, Longman Cheshire.

Queensland Department of Natural Resources. 1991, Strategy for Integrated Catchment Management. Brisbane, Queensland Department of Natural Resources.

Rogers, P. 1993, Integrated urban water resources management. Natural Resources Forum 17[1], 33–42.

Syme, G. J. 1995, The PRIME procedure: An overview. Making Catchment Management Happen, 30 May–1 June 1995 Gunnedah NSW, Proceedings. 61–63. Liverpool Plains Land Management Committee.

White, G. F. 1997, The river as a system: A geographer's view of promising approaches. Water International 22[2], 79–81.

2

River basin governance: experiences and evaluations

2.1 A TYPOLOGY OF RIVER BASIN ORGANISATIONS

2.1.1 Perceptions of the 'River Basin'

The idea of the 'river basin' as a resource unit of management is not new. Rivers and their source basins have been the site of human occupation and a focus of management for many years, since the earliest riverine civilisations: the Nile, the Yangtze, the Indus, the Tigris and the Euphrates river civilisations depended on large rivers for potable water supplies and for some, irrigation water. Early Egyptian river culture was synchronised to recurring floods and developed a cultural and religious relationship with the river. Several millennia later, this same iconic river was harnessed for power production and irrigation water in the Aswan Dam scheme. More recently, it was the focus of large scale river basin management, with the Nile Basin Initiative, which takes a whole of basin approach to natural resources management.

The river basin was recognised in 1792 by Philippe Bauch in presenting a memoir to the French Academy of Sciences in which he used the river basin as a topographical unit of management (Smith 1969). Mapping the 'river basin' became established in the late 18th century, but it was not until the mid 19th century and the early 20th century with the rise of regional geography and its concept of 'sense of place', that the 'river basin idea' emerged as a locus of research and natural resources management practice. The use of rivers as transport routes for commerce in North America underpinned this emerging regionalism. The demand for navigation crystallised the establishment of the Rhine Commission as one of the first European experiences in river basin management.

In the 20th century, river basins were regarded as development entities. Later in this century, the integrated paradigm, as discussed in Chapter 1, emerged and today dominates river basin management thinking. Barrow (1998) promoted the term 'river basin development planning and management' (RBDPM) to encompass the synthesising post World War 2 development initiatives in river basin management with integrated river basin management (IRBM). He sees RBDPM as

> encompassing activities which, although within the drainage basin, can be distant from river channels and may involve resources other than or in addition to river water. Whether national or international, RBDPM involves three main activities: planning, management and conflict resolution. It is much more than water resources development, seeking to integrate three interrelated, but separately evolved concepts: (a) multipurpose development; (b) an integrative role of the drainage basin unit; (c) the acceptance of intervention to promote economic development (typically seen to be improvement of social welfare or regional conditions). (p. 171)

This definition differs from IRBM as defined in Chapter 1, and illustrates three different river basin conceptualisations: resource development, natural resources management and ecosystem preservation. In both developed economies and transitional, developing economies of the world today, river basin management vacillated in importance as a natural resources management imperative during the late 20th century. Very recent approaches, spearheaded by international fora and international water NGOs, have established IRBM (Global Water Partnership Technical Advisory Committee 2000; Jonch-Clausen and Fugl 2001; Global Water Partnership 2002) and integrated approaches to the governance of water (Rogers and Hall 2003). In this context, IRBM is a subset of integrated water resources management.

The Global Water Partnership, in its ToolBox on integrated water resources management, suggests that river basin organisations (RBOs) are specialised organisations set up by political authorities, or in response to stakeholder demands. RBOs deal with the water resource management issues in a river basin, a lake basin or across an important aquifer. RBOs provide a mechanism for ensuring that landuse and needs are reflected in water management and vice-versa (Global Water Partnership 2002) (Tool B 1.04).

While changing perceptions of rivers and their associated basins are not new, there is no doubt that the governance of river basins is changing. Some differing, and challenging, conceptualisations of river basins are:

- *Basins as hydro-ecological units.* River basins are hydrological units of the earth's surface over which water runs to a lowest point (estuary, lake, delta). A river basin is considered as a physical and ecological whole, dependent on water for ecosystem functions and a separate entity on the earth's surface. It thus makes sense to see it as a planning and management unit for natural resources management. Furthermore, river basins are integrated ecological systems, with the management responses recognising ecosystem interdependencies. This approach developed in the 1980s and 1990s is apparent today and reflected in new initiatives for sustainable river basin management.

- *Basins as economic production systems.* In some river basins, navigation, hydroelectric power production and irrigation water have been and continue to be the perceived triumvirate for economic growth. Water resources are seen as the mechanism for national economic development, especially in developing countries. This mechanism is also regarded as a fundamental for sustainability too, so that 'development trajectories' can be visualised for future generations—water is a key for human survival and ensuring human well being (Shah, Molden and Sakthivadivel 2003). This approach recognises the constraints of the available water resources systems within river basins, and where these constraints limit growth, inter-basin transfer and/or expanded conjunctive use of groundwater resources are seen as a solution.
- *Basins as places of peoples and cultures.* The perception of the majority of basin residents is probably not one of living in a river basin, rather they see their location in terms of other factors: closeness to work, being near family, access to goods and services, part of a neighbourhood and many other reasons. This lack of river basin awareness presents significant challenges to river basin managers. The response by some RBOs is river basin education as a means of raising the awareness of basin residents, their use and impact on the river basin's natural resources. However, as the contribution of the individual is frequently miniscule, residents' awareness fades and education campaigns need to be ongoing.
- *Basins as landscapes of stakeholders' decisions.* The importance of these perceptions is less significant when it comes to stakeholders who have a more direct involvement in water use of the river basin, for example irrigation farmers, hydro-power authorities, water-dependent industries and urban and rural water service providers. They have a greater vested interest in the water resources of a basin than citizens of the river basin and are the prime decision-makers. This issue of decision making (governance) is discussed further in Chapter 3.
- *Basins as the locus of common pool natural resources management.* In many ways, a river basin's land and water resources are common goods. Since Hardin's first conceptualisation of the 'tragedy of natural resources commons' (because of the lack of collaborative management), much work has been done by Ostrom to define institutions to deal with commons management (see also Chapter 7) (Harden 1968; Ostrom 1971; Ostrom 1990; Dietz, Ostrom and Stern 2003). A river basin is a 'commons' dilemma—the problems are everyone's but no-one's. The basin is owned by all (as both private or public property but really the resources of the State) yet never effectively managed by all in any form of governance.

2.1.2 A Taxonomy of River Basin Organisations

RBOs can be seen as the solution to the commons dilemma. The role and functions of RBOs around the world vary significantly and there are many basin organisation experiences. Rogers (1993), for example, recognizes over 280 international river basin bodies. RBOs can be classified as those occurring entirely within a country, such as the Krishna in India, the Delaware in the USA, the Murray–Darling in Australia and the Yangtze in China. RBOs can also be developed to address basins which straddle international borders. International river basins have now been mapped and classified. Annex 1 provides a taxonomy of the international river basins of five continents: Africa, Asia, Europe, North America and South America.

The functions of RBO vary: water allocation, resource management and planning, education of basin communities, natural resources management strategies and programs of remediation of degraded lands and waterways, and others. They may also play a role in consensus building, facilitation and conflict management. RBOs are the organisations with an integrated function over a delineated area of land (the basin) for improved land and water governance. This area can cross international, state and local government boundaries and thus presents significant administrative, political and cultural challenges.

There are many types of RBOs, including authorities, trusts, commissions, committees and others. The roles and responsibilities of RBOs have changed through time. They include monitoring, investigating and coordinating river committees (oversee conditions and trends in the use and quality of basin resources and suggest methods to coordinate management for improved governance); planning and management commissions (more prescriptive than the first) and development and regulation authorities (regulatory bodies and enforcement agencies) (Millington 1999). Some of the earliest RBOs (such as the Tennessee Valley Authority) used top–down methods of multi-objective planning for poverty reduction, navigation, soil conservation, flood management. More recent examples (such as Murray–Darling Basin Commission, Australia) use engagement processes with basin communities to harness ownership and incorporate stakeholders in integrated approaches to natural resources management for salinity management and reallocating water resources.

Barrow (1998) suggests the proliferation of river basin experiences reflects RBOs' changing purpose through time, and calls for an inclusive term: river basin development, planning & management (RBDPM) which involves planning, management and conflict resolution. He suggests that basin-wise water resources management has both a conservation and development orientation:

> integrated and comprehensive approaches share: (i) adoption of basin-wide program; (ii) multi-purpose development; (iii) a comprehensive regional development goal – some see RBDPM as an improvement of the integrated approach with a welfare focus ... it makes sense to use the term 'integrated' for an approach which goes further than 'comprehensive'... so that water can be a tool for social and economic development.

The form and role of an RBO is closely linked to its historical and social context. There is no 'right' model when choosing from the nine types provided in the taxonomy below. The following typology is an illustrative, non exhaustive sample of RBOs and river basin management experiences in different river basin settings and were drawn from Barrow (1998), Burchi (1985), Downs, Gregory and Brookes (1991), Hooper (1995), Sturgess (1997), Duhaim Law: Online Resources and Legal

Dictionary (http://www.duhaime.org/, accessed November, 2004), and websites accessed 2003 to 2004, and the author's own fieldwork experiences, 1996–2004.

Type 1: Advisory committee

A formalised or quasi-formal organisation in which individuals take responsibility for undertaking action planning and provide advice; governments 'hand over' strategic planning to such organisations; they frequently have no or limited legal jurisdiction.

Example 2.1 Dawson Catchment Coordinating Association

Country/Countries: State of Queensland, Australia	Basin size: 50,800 km^2
Core functions: The Dawson Catchment Coordinating Association (DCCC) evolved from the Dawson Development Association and was instituted in 1998. The DCCC is a civic society organisation which receives funding from national and state government grants, and local in-kind contributions. It fosters coordination, promotes understanding and integration of natural resources management matters, promotes the development & coordination of Landcare projects (remedial catchment works undertaken by local action groups), and serves as a forum for community and government discussions on catchment management. In 1998, following widespread consultation, it released a *Dawson River Catchment Strategy*—a framework for partnership development between industry, community and government to enact natural resources management; created within the Integrated Catchment Management Strategy of Queensland. *Estimated population*: 31,000.	
References and websites: Bent (1998); http://www.dawsoninfo.org/plan/plan.html	

Example 2.2 Verde Watershed Association

Country/Countries: Arizona, USA	Basin size: 17,000 km^2
Core functions: Established in 1992 out of a conference convened to discuss resource management in the basin, it called for environmental flows in the Verde Basin and submitted an initiative to President Clinton's American Heritage Rivers Initiative; the Verde Watershed Association (VWA) represents progress in efforts to resolve watershed issues without enactment of new laws or reliance on litigation; the organisation was founded on the belief that effective river management and protection schemes can only succeed with local consensus and support; it acts as an advocate for watershed management issues.	
References and websites: www.vwa.org	

Type 2: Authority

An organisation which makes planning decisions at a central or regional government level; may set and enact regulations, or have development consent authority;

authorities are founded on democratic principles and a framework of law to which all relevant individuals and institutions are subject in a basin setting.

Example 2.3 Grand River Conservation Authority

Country/Countries: Province of Ontario, Canada	Basin size: 7000 km^2
Formed in 1948, the Grand River Conservation Authority (GRCA) operates at the provincial level with cooperation of municipal authorities and has used simulation models of point and non-point source flows to address water pollution problems in a rapidly urbanising catchment. It coordinates municipal action and can be considered a federation (see below). The core functions of the GRCA are conservation and restoration of natural resources, monitoring and advice on urban development, and management of natural resources other than gas, oil, coal and minerals. It has successfully used community participation strategies to engage greater ownership of catchment scale decisions, particularly by local municipalities. The authority has a powerful information facility (GIS) which displays regional scale information to facilitate decision-making.	
References and websites: Shrubsole, Hammond and Green (1994), Heathcote (1998); http://www.grandriver.ca/	

Example 2.4 Niger Basin Authority

Country/Countries: Benin, Burkina, Cameroon, Chad, Cote d'Ivoire, Guinea, Mali, Niger, Nigeria	Basin size: 2,273,946 km^2
Core functions: Although titled an Authority, the Niger Basin Authority (NBA) functions primarily as an organisation to promote cooperation. It does this among the member countries to ensure integrated development in all fields through development of its resources, notably in the fields of energy, water resources, agriculture, forestry exploitation, transport and communications, and industry.	
References and websites: http://www.abn.ne/index.html	

Example 2.5 Tennessee Valley Authority

Country/Countries: USA	Basin size: 106,000 km^2
One of the oldest and best-known RBOs, the Tennessee Valley Authority has emerged from being a multi-purpose river basin management organisation to a power utility. It was originally established for the purposes of navigation improvement, soil conservation, flood management and poverty alleviation in the Tennessee Valley. Its core functions now include a broad mandate of sustainable economic development, tied to supplying power and managing a river system. Management responsibilities include: minimising flood risk, maintaining navigation, providing recreational opportunities and protecting water quality. *Estimated basin population*: 4.5 millions.	
References and websites: www.tva.gov	

Type 3: Association

Similar to an advisory committee, this is an organisation of like-minded individuals and groups with a common interest. In a river basin they have varying roles: providing advice, stimulating basin awareness, education and ownership of basin natural resources management issues; educational functions and information exchange.

Example 2.6 Missouri River Basin Association

Country/Countries: USA	Basin size: 1,372,700 km^2
Core functions: The Missouri River Basin Association (MRBA) acts as an information exchange organisation, promoting the sustainable use of the Missouri River basin's natural resources. It advocates coordinated management of flood control, aids to navigation, irrigation, power generation, municipal and industrial water supplies, stream-pollution abatement, sediment control, preservation and enhancement of fish and wildlife and the creation of recreation opportunities. The MRBA has advocated adaptive management for the management of the Missouri River system.	
References and websites: http://www.mrba-missouri-river.com/; http://www.usbr.gov/dataweb/html/psmbp.html	

Type 4: Commission

An organisation which is delegated to consider natural resources management matters and/or take action on those matters. A basin commission's powers vary, and include advisory/education roles, monitoring roles, undertaking works, fulfilling goals of a specific government's charter or an international agreement. Commissions normally are instituted by a formal statement of a command or injunction by government to manage land and water resources; commissions may also have regulatory powers.

Example 2.7 Delaware Basin Commission

Country/Countries: USA	Basin size: 35,000 km^2
Core functions: A clear example of an IRBM organisation, in that it coordinates river basin planning, managing, development, mediation and regulation. Programs include water quality protection, water supply allocation, regulatory review (permitting), water conservation initiatives, watershed planning, drought management, flood control and recreation. *Estimated population*: 8 millions.	
References and websites: http://www.state.nj.us/drbc/drbc.htm	

Example 2.8 Great Lakes Commission

Country/Countries: USA and Canada	Basin size: 765,990 km^2
Core functions: The Great Lakes Commission comprises states of the USA bordering the Great Lakes and associate member status of Ontario and Quebec. Its role is to promote the integrated and comprehensive development, information sharing, use and conservation of the water and related natural resources of the Great Lakes basin and St. Lawrence River. It also undertakes policy research in environmental protection, economic development and transportation. *Population served*: approximately 33.2 millions.	
References and websites: http://www.glc.org/	

Example 2.9 International Commission for the Protection of the Danube River

Country/Countries: Germany, Austria, Slovakia, Czech Republic, Slovenia, Serbia and Montenegro, Bulgaria, Moldova, Ukraine, Romania, Bosnia-Herzegovina, Croatia and Hungary	Basin size: 817,000 km^2
Core functions: Decision making, management and coordination of regional cooperation; approval of budget and annual work programmes; follow up on activities and evaluate results from Expert Groups; Joint Action Programme. The 1994 Convention on Cooperation for the Protection and Sustainable Use of the River Danube set up this Commission and the convention established mechanisms to take "all appropriate legal, administrative, and technical measures to at least maintain and improve current water quality conditions . . . of the river and its catchments . . . and prevent and reduce adverse impacts".	
References and websites: http://www.icpdr.org/	

Example 2.10 International Commission for the Protection of the Rhine

Country/Countries: Germany, France, Luxembourg, Netherlands, Switzerland	Basin size: 185,000 km^2
The International Commission for the Protection of the Rhine provides an international forum for member countries to address water quality; and provides a catalyst for increased political and public participation in dealing with pollution issues, a network of monitoring stations, knowledge exchange on best practice in water quality abatement, and a setting for the conduct of negotiations. Core functions include improvement of water quality, flood control, guaranteeing the use of Rhine water for drinking water production, improvement of the sediment quality in order to enable the use or disposal of dredged material without causing environmental harm, flood prevention and environmentally sound flood protection, and improvement of the North Sea quality in accordance with other measures aimed at the protection of this marine area.	
References and websites: Mostert et al. (1999), Dieperink (2000); http://www.iksr.org/, http://www.thewaterpage.com/rhine_main.htm	

Example 2.11 International Joint Commission, USA and Canada

Country/Countries: USA and Canada	Basin size: 765,990 km^2
Core functions: An independent binational organisation established by the Boundary Waters Treaty of 1909. Its purpose is to help prevent and resolve disputes relating to the use and quality of boundary waters and to advice Canada and the United States on related questions. The focus of this work has been addressing key water quality management programs by listing these according to deterioration in environmental condition. Through its work, some of the original 47 sites of concern have been delisted. *Population served*: approximately 33.2 millions.	
References and websites: http://www.ijc.org/	

Example 2.12 Lake Chad Commission

Country/Countries: Cameroon, Niger, Nigeria, Chad and Central African Republic	Basin size: 2,388,700 km^2
Core functions: Created in 1964, the Lake Chad Commission prepares regulations to enact its convention between members states. It collects, evaluates and disseminates information on projects in the Lake Chad Basin which are prepared by member states and recommends plans for common projects and joint research programmes. This commission has a coordinating role, keeping close contact between the contracting parties to ensure efficient utilisation of basin waters, and coordinating regional programmes; has a rule setting role regarding navigation and transport; a conflict resolution role in promoting regional cooperation; and monitors, plans and enacts national projects with regional significance.	
References and websites: http://www.cblt.org/	

Example 2.13 Mekong River Commission

Country/Countries: Vietnam, China, Thailand, Laos, Khmer PDR	Basin size: 787,000 km^2
Core functions: Promote and coordinating sustainable management and development of water and related resources for the countries' mutual benefit and the people's well-being by implementing strategic programmes and activities and providing scientific information and policy advice. The Mekong River Commission (and former Mekong River Committee) has promoted dialogue between members states for 40 years, is still implementing its vision of several lower Mekong mainstream dams, flood forecasting and warning, water balance studies, water quality monitoring, salinity control on the Mekong delta and ex-post studies of multipurpose dams in northeast Thailand. *Estimated population*: 60 millions.	
References and websites: Millington (2004), Jacobs (1999); http://www.mrcmekong.org/	

Example 2.14 Murray–Darling Basin Commission

Country/Countries: Australia	Basin size: 1,061,469 km^2
Core functions: The MDBC is a 'federation' style of commission, in that it is an agreement between five members states and a national government to jointly manage natural resources. MDBC originally established a strategy, but in 2001 developed an initiative which spells out its functions in supply water for the Murray River, and three programme areas: policy (develops and reviews basin-wide policies); knowledge generation (generate knowledge to support policies and programs) and on ground activities (improving river health, sustainable landuse, restoring riparian land systems, wetlands and floodplains and improving water quality). As well, it aims to achieve equitable, efficient and sustainable use of the land, water and environmental resources, control and/or reverse land degradation; protect and/or conserve the natural environment and conserve cultural heritage. *Estimated population*: 2 millions.	
References and websites: Millington (2004), Murray–Darling Basin Commission (2001a) and Murray–Darling Basin Commission (2001b); http://www.mdbc.gov.au/	

Example 2.15 North Carolina Environmental Management Commission

Country/Countries: USA	Basin size: Varies—several river valleys exist; representative example being Neuse River—area: 14,582 km^2
This organisation's jurisdiction is one of advocacy, coordination, education and producing guidelines for others to enact; uses a regulatory and interventionist approach, one tempered with strong citizen checks and was developed to address growing point and non-point source pollution. This commission is made up of 17 members representing lay people or those outside State Government, appointed by the Governor or by legislature. With water resources, water quality and air quality agenda, EMC enacts water quality management as a rule-making body for water quality programs (designed and implemented by County Government and supported by State agencies). This approach sets guidelines such as a 30% reduction of nitrogen levels in North Carolina rivers from 1991 to 1995 average load rating which is targeted to be achieved within 5 years. Working at the county (Local Government) level, individual resource managers (farmers, foresters, industries) have two options to manage both point and non-point source pollution: • Meet requirements by installing mandatory best management practices (these rely heavily on water control structures on farm and in towns, regulating pollution flows and redirect flows from critical areas, use buffer strips). • Join a county watershed management program in which counties and individual farmers can trade discharge permits using a marketing system to achieve country water quality requirements. In short, EMC provides a process whereby rivers are brought under a pollution regulation, in which countries develop water quality management plans which are approved or rejected by the State natural resources agency, and which are designed to meet EMC-set rules and targets.	
References and websites: http://h2o.enr.state.nc.us/admin/emc/, http://www.epa.gov/nerlesd1/land-sci/lcb/nrb/VFRDB/VFRDBmain.htm	

Example 2.16 Ohio River Water Sanitation Commission

Country/Countries: USA	Basin size: 526,000 km^2
Core functions: A basin-wide organisation with a specific water quality mandate, it operates programs to improve quality, by setting waste water discharge standards; performing biological assessments; monitoring for the chemical and physical properties of the waterways and conducting special surveys and studies. It is essentially a technical entity providing scientific information for basin-wide management.	
References and websites: http://www.orsanco.org/	

Example 2.17 Tarim Basin Water Resources Commission

Country/Countries: China	Basin size: 200,000 km^2
Core functions: Irrigated agriculture development and improvement; establish mechanisms for sustainable use, development and management of water resources in the Tarim Basin and partially restore and preserve the 'green corridor' in the lower reaches of the Tarim River. The commission is also responsible for planning and developing policy. Millington (2004) cites the Tarim Basin Water Resources Commission as an RBO which has developed basin-wide water allocation procedures: it determines desired river health levels in lower reaches of the basin, by specifying sub-basin, upstream surface water flows as contributions to this desired level. *Estimated basin population*: 5 millions.	
References and websites: Millington (2004); http://www.worldbank.org.cn/English/content/693g1207183.shtml	

Type 5: Council

A formal group of experts, government ministers, politicians, NGOs and lay people brought together on a regular basis to debate matters within their sphere of basin management expertise, and with advisory powers to government. A council is contrasted with a commission which, although also a body of experts, is typically given regulatory powers in addition to a role as advisor to the government.

Example 2.18 Fraser Basin Council

Country/Countries: Canada	Basin size: 240,000 km^2
Core functions: Commenced in 1997, the Fraser Basin Council (FBC) has a broad mandate of sustainable development, focusing on pollution control, environmental improvement and integrated basin management. The FBC is an advisory organisation and facilitates actions across its basin area by bringing together stakeholders and formulating joint action programs. Examples include: • Review of the Drinking Water Protection Act and made recommendations to government on the completeness, effectiveness and efficiency of the Act. • Lead role in the Joint Program Committee of 36 federal, provincial and local government agencies working towards Integrated Flood Hazard Management. *Estimated population*: 2.7 millions.	
References and websites: http://www.fraserbasin.bc.ca/	

Type 6: Corporation

A legal entity, created by legislation, which permits a group of people, as shareholders (for-profit companies) or members (non-profit companies), to create an organisation which can then focus on pursuing set objectives, and empowered with legal rights which are usually only reserved for individuals, such as to sue and be sued, own property, hire employees or loan and borrow money. Also known as a 'company'. The primary advantage of a for-profit corporation is that it provides its shareholders with a right to participate in the profits (by dividends) without any personal liability, because the company absorbs the entire liability of the organisation.

Example 2.19 Damodar Valley Corporation

Country/Countries: India	Basin size: 24,235 km^2
Core functions: Began in 1948, the Damodar Valley Corporation focuses on integrated area development: promoting and operating schemes for irrigation, water supply and drainage, generation of electrical energy, flood control, navigation, reforestation and controls of soil erosion, and public health and the agricultural, industrial development.	
References and websites: Saha (1979); http://www.dvcindia.org/index.htm	

Example 2.20 Snowy Mountains Engineering Corporation (now Snowy Hydro)

Country/Countries: Australia (interstate)	Basin size: km^2 (catchments of the Murrumbidgee, Snowy, Murray and Tumut Rivers)
Core functions: Formally, the Snowy Mountains Engineering Authority, established in 1949 as an act of national parliament, to establish the authority and construct the Snow Mountains Scheme, one of Australia's hydro-electric and irrigation development schemes. On completion in 1974, the Scheme consisted of seven power stations, 16 major dams and 80 km of aqueducts. On June 28, 2002, the Snowy Mountains Hydro-electric Authority was corporatised. This resulted in the merge of the Authority and Snowy Hydro Trading P/L (established to trade electricity into the national grid) to become Snowy Hydro Limited. Built in the national interest with the support of the New South Wales, Victorian, South Australian and Commonwealth Governments, the Scheme today provides electricity (3756 MW capacity) to the National Electricity Market and drought security to Australia's dry inland. Snowy Hydro Limited operates and maintains the Snowy Mountains Scheme. *Current mission statement*: "To deliver superior financial returns by being the preferred flexible supplier of energy and related products; developing our people and utilizing our physical assets and water resources in order to exceed customer and stakeholder expectations while demonstrating best practice in safety and health, asset and environmental management."	
References and websites: www.snowyhydro.com.au	

Type 7: Tribunal

A basin entity which has formalised procedures and quasi-judicial powers; a heavy emphasis on bureaucratic decision making; stakeholders may formally participate through hearings; major decisions are taken by independent bodies, like a water pricing tribunal. A tribunal acts as a special court outside the civil and criminal judicial system that examines special problems and makes judgements, e.g. a water tribunal, which resolves disputes between water users. Very few such entities exist purely for river basins management purposes but rather for special purposes, e.g. government pricing tribunals. Some tribunals have specific water functions which are a component of a broader river basins management process, where an RBO may or may not exist. These entities have essentially no traditional powers of civil government and do not report to other government agencies, except where a local government body may oversee. They play an important role in developed countries and many developing countries.

Example 2.21 Valencia Water Court

<table>
<tr><td>Country/Countries: Spain</td><td>Basin size: Several</td></tr>
<tr><td colspan="2">Core functions: This very old institution, perhaps one of the most long-lasting in the history of water institutions is a local entity which resolves disputes amongst water users. It applies mainly to groundwater basins in the Valencia region. Similar Arabian water organisations were established some 1000 years ago. They provide a water rights system that is tied to land and is not able to be sold separately, operating rules, equitable full cost recovery from customers and an internal enforcement mechanism. It is believed that the Valencia Water Court has resolved water disputes for over 300 years.
Llamas 2003 states that

. . . Spain has a long tradition of collective management of common pool resources. Probably the Tribunal de las Aguas de Valencia . . . is the most famous example. This Court has been meeting at noon every Thursday for many centuries at the entrance of Cathedral of Valencia to solve all the claims among the water users of a surface irrigation system located close to Valencia. All the members of the Court are also farmers. The decisions or sentences are oral and can not be appealed to a higher court.</td></tr>
<tr><td colspan="2">References and websites: (Llamas 2003) Van Ittersum and Van Steenbergen (2003), Tardieu (2004)</td></tr>
</table>

Type 8: Trust

Trust is a legal device used to set aside money or property of one person for the benefit of one or more persons or organisations. It is an organisation which undertakes river basin works, develops and implements a strategic plan, its mandate is to be the river basin 'advocate', it coordinates local programs through Memoranda of Understanding or other agreements, it raises local levies (funds) for is works and programs. A Trust keeps monies raised in 'trust' for the benefits of its citizens.

Example 2.22 Hawkesbury–Nepean Catchment Management Trust (now part of the Hawkesbury–Nepean Catchment Authority (HNCMA)

Country/Countries: Australia	Basin size: 22,000 km^2
Core functions: Deliver a healthy, productive and diverse Hawkesbury–Nepean River system and catchment to present and future generations. The Hawkesbury–Nepean Catchment Management Trust worked to build and strengthen partnerships between government, business and the community to enable them to work together to protect and restore the catchment. This trust has been dismantled. The HNCMA specifies river basin management through setting water quantity and quality guidelines, similar to catchment authorities elsewhere in the state. This is done by specifying interim flow objectives for rivers in its basin.	
References and websites: http://www.hn.cma.nsw.gov.au	

Type 9: Federations

A collaboration of departments within one government or between state and national governments to establish and undertake actions for river basin management. Local government groupings have emerged in some locations such as in the USA for regional natural resources governance. Governance actions at various levels (national, state and local) include agreements on water sharing and water quality management, shared statements of intent, shared policy development, information exchange, joint actions for management of ecosystem degradation. Collaboration is expressed in terms of framework directives, cost-sharing arrangements, joint statements of intent, partnerships, joint programs and agreed policy.

Example 2.23 Chesapeake Bay Commission and the Chesapeake Bay Agreement

Country/Countries: USA	Basin size: 165,760 km^2
Core functions: Chesapeake Bay Commission (CBC) is a tri-state legislative commission created in 1980 and operates as a federation of state government departments. The CBC serves as the legislative arm of the multi-jurisdictional Chesapeake Bay Program and acts in an advisory capacity to their respective General Assemblies. Its program is the unique regional partnership that directs and conducts restoration of the Chesapeake Bay, since the signing of the 1983 Chesapeake Bay Agreement which brings together the States Maryland, Pennsylvania, Virginia and the District of Columbia, with the CBC, the EPA (representing the federal government) and the participating advisory groups. Initial research (1970s) pinpointed three areas requiring immediate attention: nutrient over-enrichment, dwindling underwater Bay grasses and toxic pollution and the Bay Program evolved as the means to restore this resource. The Chesapeake Bay Agreement sets goals to reduce the nutrients, nitrogen and phosphorous, entering the Bay by 40% by 2000. Achieving a 40% nutrient reduction will ultimately improve the oxygen levels in the Bay waters and encourage aquatic life to flourish. In 1992, the Bay Program partners agreed to continue the 40% reduction goal beyond 2000 as well as to attack nutrients at their source—upstream in the Bay's tributaries. As a result, Pennsylvania, Maryland,	

Virginia and the District of Columbia began developing tributary strategies to achieve nutrient reduction targets. On June 28, 2000, the Chesapeake Bay Program partners signed the new Chesapeake Agreement which will guide the next decade of restoration and protection efforts throughout the Bay watershed.
References and websites: http://www.chesapeakebay.net/index_cbp.cfm

Example 2.24 Council of Great Lakes Governors and the Great Lakes Basin Water Resources Compact

Country/Countries: USA and Canada	Basin size: 765,990 km^2
Core functions: The Council of Great Lakes Governors is a partnership of the Governors of the eight Great Lakes States—Illinois, Indiana, Michigan, Minnesota, New York, Ohio, Pennsylvania and Wisconsin. It was formed in 1983 to tackle the severe environmental and economic challenges of the Great lakes Basin. In recent years, the Canadian Premiers of Ontario and Quebec have joined the Council Governors to advance economic development of the Great Lakes region. The Great Lakes contain one-fifth of the world's fresh water and the region's companies manufacture 60% of the continent's steel and 60% of automobiles made in North America. The Great Lakes Governors have made fundamental and sustainable changes in areas such as education, welfare reform, trade and landuse management. The mission of the Council is to encourage and facilitate environmentally responsible economic growth, by establishing a cooperative effort between the public and private sectors. The Council works directly for the eight Great Lakes Governors on projects, such as water management, ballast water initiative, the Great Lakes Trade Initiative and others. Agreements are used to update the way the Great Lakes and the waters of the Great Lakes Basin are managed and protected. The 2004 water compact amongst the states builds on the Boundary Waters Treaty of 1909, Great Lakes Water Quality Agreement of 1978, Great Lakes Charter of 1985 and Great Lakes Charter Annex of 2001. *Population served*: approximately 33.2 millions.	
References and websites: www.cglg.org	

Example 2.25 European Commission—Directive on River Basin Management

Country/Countries: All European Union member states	Basin size: All of EU area Sample basins: Rhine—172,900 km^2, Danube—790,000 km^2, Guadiana—67,900 km^2, Oder/Odra—122,400 km^2, Rhone—100,200 km^2
Core functions: The European Commission is the entity responsible for the enactment of the European Water Framework. Within this framework, the Directive on River Basin Management applies to river basins throughout the EU, therefore encompasses major and smaller river basins such as the Rhine, Danube, Guadiana, Oder and Rhone. In this way, it is an initiative rather than the specification of one type of RBO, and it applies to all member states of a 'federation' (the European Union). The Directive requires development of river basin plans. The initiative is built on the policy provisions of preventative action, the precautionary principle and the polluter pays principles. The	

overall aim is to maintain and improve the quality of the aquatic environment, in both national and international settings. Member states will work out their own form of entity to enable river basin plans.
References and websites: Chave (2001); http://europa.eu.int, http://europa.eu.int/comm/environment/water/water-framework/overview.html

Example 2.26 Sand River Integrated Catchment Management Project

Country/Countries: South Africa	Basin size: 7096 km^2
Core functions: The Sand River project is a national pilot project for Integrated Catchment Management (ICM), which emerged in 1996 as a national government initiative. The ICM approach allows clear segmentation of river systems into functional management units (catchments and sub-catchments) which can then be linked together to form an overall management plan for an entire river basin. The management units should encompass linkages between components and will usually consist of the whole catchment or a similar geographical unit, such as a sub-catchment (Department of Water Affairs and Forestry 1996). The Sand River project is an example of the South African Government's Integrated Catchment Management Strategy. It is lead by AWARD, a non-profit organisation that promotes and supports equitable and accessible community water services and resources in the catchment. The thrust of the project is to facilitate stakeholder participation of stakeholders including: local government, traditional leaders, water committees, Department of Water Affairs and Forestry, extension officers, the Bushbuckridge Water Board, Community Development Fora and schools. One of the major elements of the project is the development and implementation of a methodology that creates the capacity of all stakeholders to actively participate in natural resources management. The project expects to develop and implement river basin management plans in a truly participatory way. *Estimated population*: 0.4 millions.	
References and websites: http://www.iwmi.cgiar.org/dialogue/files/Dialogues/Africa/INFOSHEET_SA_ savetheSand.htm	

2.2 EVALUATING IRBM EXPERIENCE

2.2.1 River basin governance models

In this section, we examine the evaluation of different river basin management experiences. Before the evaluations, it is important to recognise how governance of IRBM has been constructed by administrations to deal with the basin commons dilemma. These constructs are interpretations of governments' role as the top down instigator of basin management. Governments and their public administrations continue to play a leading role in the formation of river basin initiatives.

Synnott (1991) used three models to demonstrate how integrated approaches to river basin governance have been formulated. The first, the Economic Model, recognises river basin management problems as largely a problem of poorly defined property rights. Because everyone, and therefore no one owns the river basin, little effective action is taken to solve resource management problems, and there is little incentive for individuals to care about third party effects of poor land management. The issue of land property rights is fraught with ethical and legal problems, but there has been some movement towards water property rights, and this could lead to the right of individuals to discharge a specific level of pollution or buy additional rights to do so. However, little real effort for integrated land and water management has occurred with this approach as it relates to river basin planning or catchment management.

The second model Synnott described was the *Government Intervention Model*. Policy objectives and priorities are set and then appropriate implementation mechanisms are developed. Intervention occurs through institutional reform, for example changed policy to adopt integrated approaches. This model has been generally resisted as an appropriate mechanism, rather campaigns focussing on individual and community participation are used. The Government Intervention Model generally avoids difficult decisions and long-term financial commitment by government. It has less attraction to government as governments prefer to pass on resource management decisions to user communities.

The third model is the *Social Response Model*. This involves no direct government intervention, and tends to be an issue driven/threat response process. It encourages awareness and research to find solutions. It uses normative persuasion rather than prescriptive direction. The value of this model is that

- it does not challenge the autonomy and rights of individual property owners;
- it does not commit government to long-term expenditure obligations;
- it has a high profile for government in its execution and
- it has widespread effects.

However, real long-term durable gains are likely to be small using the Social Response Model, because it produces unintended responses from individual landowners. The majority of rural landowners often see resource management problems as 'out there' problems, not as their problem and fail to take ownership of them. This negates the prescriptive requirement of the Social Response Model which is built on local ownership of resource management. Synnott maintained that this is one of several issues restricting the implementation of Social Response Models for IRBM. Others relate to contrasting private and public interests operating in any resource management problem (Table 2.1), which complicate management responses.

These dualisms imply that, in any river basin management situation, there will always be winners and losers, and the conflicts they generate cannot be resolved on an individual, local basis.

Synnott (1991) applied his perspectives to Australia. He maintained that national and State governments (implied by their being the representative implementation agencies) had remained locked into a Social Response Model approach to implement integrated catchment management (ICM). This was echoed some 4 years later in another Australian study. Early implementation of IRBM in Australia was

Table 2.1 Dual perspectives in natural resources management at the river basin scale

Private interests	Versus	Public interests
Urban interests/costs	Versus	Rural interests/costs
Local issues/management	Versus	Regional issues/management
Present	Versus	Future
Local	Versus	Regional
Risk acceptance	Versus	Risk avoidance
Coercion	Versus	Cooperation
Strategic decision-making	Versus	Ad hoc decision-making

Source: Modified from Synnott (1991).

due to the lack of resources and government procrastination, and an over-reliance on the Social Response Model, although adoption of the integrated philosophy was well advanced (AACM International and Centre for Water Policy Research 1995). Mitchell and Hollick (1993) likewise found in Western Australia that there was the acceptance of the philosophy of integrated approaches, the recognition of the value of the products (clean water, guidelines for action, basin management plans) but considerable uncertainty of the process.

Together with Canada, Australia was one of the first countries in the world to further the conceptual development and practice of integrated land and water management in the 1980s (Burton 1984; Burton 1985; Burton 1986; Mitchell 1987; Murray–Darling Basin Ministerial Council 1987; Burton 1988; Blackmore 1995). The initial experience used a social response approach with restricted use of direct government intervention, seen in limited interest in the use of economic instruments and regulatory mechanisms to enforce ICM. ICM, as it has become known in Australia, however, was institutionalised with the passing of a Catchment Management Act in 1989 in New South Wales and the formation of the Murray–Darling Basin Commission following the Murray–Darling Basin Agreement of 1985 (Murray–Darling Basin Commission 2001a). However, there is the need, after nearly 20 years of experience, for long-term commitment to coordinating structures at the basin scale, beyond that established in the Murray–Darling Basin. There is still a limited national landuse policy, but there is some development of national standards for water resource management (as provided in the National Water Quality Management Strategy).

There is also the complication of an ever-present turbulent institutional environment and resulting organisational change in national and state water and land management agencies. These agencies evolved through successive national and state administrations into a user pays/corporatised environment with smaller government and more owner-management of natural resources. There has also been many changes to the functions of resource management agencies, with the separation of powers of resource management, regulation and service provision. The positive outcome of this process is that the integrated basin/catchment approach is widely endorsed and practised in Australia.

2.2.2 Evaluations of river basin management performance

While integration of land and water resources management is widely used, it remains difficult to accomplish in practice, from the local watershed to the basin scale (Born and Sonzogni 1995). The critical integration issues include the need for strengthened *coordination* mechanisms, role specification, clarity in the divesting of government powers and responsibilities, financing IRBM (who pays?), community involvement, the need and scope of integration, leadership skills, information access and exchange, links with regional and local planning, bureaucratic resistance and property rights problems (Siann 1981; Mitchell 1983; Mitchell 1986; Environmental Protection Authority 1993; AACM International and Centre for Water Policy Research 1995; Born and Sonzogni 1995; Environmental Protection Agency 1995; Barrow 1998; Heathcote 1998; Born and Genskow 1999; Bellamy et al. 1999; Hooper, McDonald and Mitchell 1999; Bellamy and Dale 2000; Margerum and Born 2000; Bellamy et al. 2001). Some of these problems are encapsulated in Table 2.2.

There have been increasing research and consultancy activities in the evaluation of the effectiveness of RBOs. Mitchell and Hollick (1993) were amongst the first to conceptualise the idea of ICM and analyse its use in large Western Australian catchments. They recognised that ICM was an approach that required strategic, systems-based and participatory functions to be used at river basin scales. Bellamy and Dale (2000), AACM International and Centre for Water Policy Research (1995) and Synnott (1991) undertook reviews of the effectiveness of Australian catchment management and found that a range of governance issues precluded the effectives of catchment management, including the lack of coordination of economic development and conservation planning at the river basin scale. Hooper, McDonald and Mitchell (1999) citing two decades of experience in the USA, Canada and Australia, identified ten causes of disillusionment and five recommendations for enhancing integrated approaches to resource management. Their study also found that governance issues prevailed. Born and Sonzogni (1995) demonstrated the use of local integrated land and water management in Dane County, Wisconsin's Black Earth Creek watershed. New analytical tools (such as GIS), thinking strategically

Table 2.2 Key factors that influence watershed management initiatives

Exogenous factors	Endogenous factors
Nature of the ecological setting and related use problems	Partnership initiation
	Composition
Demographic and socio-economic setting	Statement/clarity of purpose
	Leadership
Situational history	Staffing
Issue salience	Governmental commitment and support
Regulatory/programmatic context	Funding
	Watershed plans

Source: From Born and Genskow (2000b).

by the Association, strategizing actions, accessing a new state stewardship program for acquisition of critical riparian lands and regularised interactions between stakeholders all worked to engender and implement integration action. Hooper (2001) showed that the failure of administrative stability, lack of ongoing funding, identity and confusion over roles precluded the lack of an RBO's capability to exchange information for planning purposes. Many studies show repeatedly that inter-organisational collaboration is a major stumbling block to IRBM and smaller scale catchment management (Margerum 1995; Kenney 1997; Margerum and Born 2000; Born and Genskow 2000a; Born and Genskow 2001; Jaspers 2003; Healthy Rivers Commission of New South Wales 2003; Hillman, Aplin and Brierley 2003; Imperial 2004; Cole, Feather and Munch 2005).

Issues such as different perceptions, conflicting missions and objectives of catchment groups, constant staff turnover, lack of continuity and consistency in human resources, transferability of continuity of management skills negatively characterise catchment management in Australia today. AACM International and Centre for Water Policy Research (1995), for example, found in an Australian setting that

- the current organisational structure for catchment management was questionable because of the vested interests of participants;
- the current institutions for catchment management were weak and uncertain. The cost and benefits of these institutions are unclear;
- catchment management decision-making was episodic and responsive to political, local and environmental pressures and
- there was a need to empower catchment management committees and this would require changing constitutions and acts of State government. This would provide these organisations with a statutory basis, and also checks and balances for accountability and control if the legislation was properly developed.

Lang (1986a) and Lang (1986b) maintained that, in Canada at least, the use of an integrated approach tended to be more rhetoric than reality. He suggested that the lack of application of an integrated approach was due to

- the lack of development of concepts and methodology;
- the lack of awareness by practitioners of others' successes and failures in implementing an IRBM approach and
- the absence of a focus for resource management professional development: such as a professional association, a journal, regular conferences, and active networking among practitioners.

These problems have weakened the more effective implementation of IRBM in Canada.

Overall, these studies found that coordination was and remains a fundamental problem and that mechanisms for integrated decision-making across sectors and regions were the paramount constraints to IRBM. In a major international study through many continents and basin experiences Barrow (1998) encapsulates this conundrum in river basin governance:

> It is difficult to overcome the problems of complexity and find a practical framework to better integrate environmental, socioeconomic and policy issues at a regional scale. (p. 180)

To address this conundrum, one first needs to specify the problems and their causes, then suggest solutions. Several common, critical problems are listed in Table 2.3. These were derived from the above studies, from Barrow (1998), Hooper, McDonald and Mitchell (1999), Le Moigne et al. (1994), Shah, Molden and Sakthivadivel (2003) and from the author's numerous discussions in river basin

Table 2.3 River basin management problem analysis

Problem	Possible causes	Remedy
Inflexible planning process and overambitious in what is sought	High ideals of constituents; ignorance of resource use potential and constraints	Use a stepped approach; do what is achievable first
Lack of baseline data and monitoring	Scarcity of funds, trained personnel, institutional difficulties, harsh environment	Better simulation modelling
Assumptions	Planning is based on false assumptions	Better supervision of the planning process; put an adaptable and flexible planning processes in place
Planners and managers attempt to solve complex, 'wicked' problems by using one-dimensional solutions	Lack of interdisciplinary training, inability of managers and leaders to think across disciplinary boundaries	Sensitive, multidisciplinary study adapted to local and or regional needs; flexible adaptive approaches to cope with unforeseen problems; community participation to try and ensure administrators are accountable and heed people's needs
Ignoring downstream impacts of upstream activities—produces inequitable outcomes	Decision-making across the basin is not coordinated and little communication between key decision processes	Need for better integration of traditional users with current and proposed RBM activities
BOOT (build-operate-oversee-transfer) method of development precludes local ownership	The legal permitting and environmental impact assessment procedures (if used) do not permit input from constituents beyond those in the planning process (often dominated by government development agencies and contracted construction companies)	Need for local training and capacity building during construction phase of project development; improved use of environmental impact assessment

Table 2.3 *Continued*

Problem	Possible causes	Remedy
Negative economic and social impacts of river basin management and development planning	Watered-down or sidestepped environmental impact assessment of basin development activities	Accountability procedures for RBM organisation's plans; integrate impact assessment with planning procedures
Developing country priorities differ from those of developed countries (see also Table 2.4)	A large informal water sector exists; tropical and subtropical/monsoonal hydrology differs from temperate region hydrology; managing rainfall where it falls is the key to water management rather than 'managed' water; institutional change does not comes through 'western' models of institutional reform	River basin management plans driven by bottom up water sector initiatives with strong NGO and village level management; water harvesting in wet seasons and efficient storage for dry seasons; institutional reform through stronger high level ownership of water management and using 'water champions' in government, NGOs and the informal sector; donor agencies sensitised to these approaches; capacity building of government agency staff in integrated water resources management
Lack of a power base of the RBO and failure to do more for than advise	Lack of ongoing political and administrative support; no sustained funding base	Ensure the RBO is enshrined in legislation, has ongoing funding commitment and line responsibility to the highest level of government
RBOs with power drift into bureaucracy building and corruption, once given power and financial autonomy	Lack of accountability	Ensure the RBO answers to the highest level of government; provide an independent auditing service/review process
Failure to control/manage/ influence the entire river basin area	Lack of legal jurisdiction, inability to 'tax' one end of the basin to 'pay for' problem remediation and development elsewhere	Ensure a process of equitable funding is in place and is externally arbitrated; provide a community advisory council to provide independent advice to the RBO

Continued

Table 2.3 River basin management problem analysis (*Continued*)

Problem	Possible causes	Remedy
Border conflict	RBM unable to manage cross government jurisdictions within countries and across international borders; poor communication about resource management and development issues and information	Ensure regular, frequent co-riparian exchanges of data and views; use a common language despite ethnic differences; improved ability to solve problems using conflict resolution mechanisms
Resources controlled by a range of entities and coordination too great a problem, especially at a basin scale	Poor institutional development at set up or during the development stage of the RBO	Flatter organisational structures provide better ability to coordinate than steeply hierarchical structures
Political influence	Political processes determine jurisdiction boundaries, who is engaged, funding commitments, agendas and may hijack integration procedures	Establish an independent arbitrator; enhance the visibility and power of NGOs who can influence process
Multilateral jurisdiction	Multilateral relationships impact decision-making and lead to suboptimal results	RBM body must monitor and revise implementation; need to use the RBM process to deal with multiple donor agendas
Sectoral dominance	Domination of a special interest group in an RBM entity leads to faulty outcomes	Need for open accountable planning and management

management field missions in Canada, India, USA, Australia, Spain and New Zealand.

What then is 'successful' IRBM? Opinions vary according to three fundamental perspectives: (1) the context and purpose of river basin management, (2) the degree to which integration of natural resources management, economic development and conservation has evolved and (3) the stage of development of river basin governance as an integrated decision-making process, both in RBOs (if they exist) or initiatives for river basin management across different jurisdictions. The success of IRBM is then subjective, contextual and time dependent. Furthermore, south–south exchange of experiences are more likely to produce effective implementation in river basins in developing countries, than north–south dialogues.

The remainder of this chapter provide *guiding principles* and the specification of context related *functions* which can be improve river basin management. They are the core of IRBM.

2.3 FUNCTIONS AND PRINCIPLES OF INTEGRATED RIVER BASIN GOVERNANCE

Political processes and administrative dysfunctionality play critical roles in coordinating water and land management across sectors, jurisdictions and regions. A major concern is that there appears to be an 'institutional vacuum' at the regional, river basin scale to address how traditional top–down management meets current demands for bottom–up participation. These are fundamentally *governance* issues.

River basin management is macro-scale natural resources management. It involves taking the 'big picture' of land and water resources management over large eco-hydrologically determined regions. In Section 2.2, we recognised nine types of RBOs, while Radosevich and Olson (1999) suggest basin organisations can be summarised under three broad functional types: monitoring, investigating and coordinating river basin committees; planning and management commissions and development and regulation organisations.

At more localised scales, sub-basin management plans and local government planning mechanisms cover many tasks within the context of an overall river basin management plan and specify best management options (see also Chapter 4). Returning to the macro-scale, it is often difficult to define the generic principles for the operation and function of an RBO. The functions depend on objectives of river basin management at that location and the context of existing conditions into which RBOs or river basin scale action programs have been developed or are planned. There are, however, some common principles and functions which can be applied. The following two lists, one of principles and other of functions, discuss IRBM as implemented by an RBO. The list should be used cautiously as it will require adaptation to the specific context where it is applied.

2.3.1 Principles

Principle 1: Engagement of and ownership by relevant decision-makers

River basin management design is enhanced if relevant river basin decision-makers own the process and participate under a formal, contractual arrangement, rather than ad hoc, voluntary arrangements. Australian river basin management is noted for the large distances between (often competing) decision-makers. The issue of multiple players and distant participants can be addressed by the use of Internet-based river basin information systems. Hooper (2001), however, maintained that word-of-mouth networks dominate river basin information exchange, so significant investment should be given to support meetings and face-to-face contacts.

Principle 2: Improved river basin management design

Sound scientific information guides effective IRBM. It describes resource condition and trend, the causes of resource degradation and the likely impacts of resource management options. Economic analysis and social impact assessment is used to provide ex-ante and ex-post evaluations of river basin management plans. Accurate modelling of river basin management options will enhance success, provided that modelling engages relevant river basin decision-makers throughout the process of model design, implementation and outcome review.

Principle 3: Application of diverse institutional arrangements

There are many institutional arrangements to enact IRBM. These include cost-sharing programs, tradable discharge permits and voluntary actions, as well as more regulatory practices such as environmental regulation, zoning laws and environmental standards for best practice.

Developing countries require different approaches to institutional strengthening for river basin management than those of developed countries, relating to their difference in hydrology, administrative capacity and vast numbers of stakeholders (Table 2.4). These realities suggest that the RBOs in developing countries will require methods to affect the behaviour of very large numbers of water users, methods which are not as frequently used in the far less populous models of river basin governance such as the Murray–Darling Basin Commission and the Tennessee Valley Authority.

Principle 4: Clear definition of the role of the RBO

An RBO requires a clear management role and jurisdiction which involves:

- a skills-based, independent membership of its board of directors/oversight group;
- a democratic process—members elected by the regional community;
- an accountability procedure of its management departments—reporting to an independent board of directors
- being linked to high levels of government for political influence and support;
- responsible for the core basin management business of provision of resource management planning guidelines especially to Local Government, implementation of plans of action and coordination of other agencies' actions in the river basin (these are discussed below) social and economic inventories of the river basin, management of river basin information and monitoring of outcomes.

Principle 5: Strong river basin advocacy

Successful river basin management is driven by strong leadership. Individual advocates and organisations with a strong river basin advocacy are needed to engage both willing and recalcitrant resource managers. There is always a plethora of decision-makers, multiple jurisdictions and competing players in a river basin. A river basin advocate is needed to present the case for integrated resource and

Table 2.4 Differences between developing countries' and developed countries' basin realties

Developed countries	Developing countries
Temperate climates, humid, higher river—stream density	Low rainfall, extreme climate, higher mean temperatures, lower stream density, water scarcity an emerging constraint
Population concentrated in the valleys, downstream	Densely populated in both valleys and catchment areas; population high both upstream and downstream of dams
Water rights based on riparian doctrine and prior appropriation	Water rights based on rights to rainfall or ground-water; people's notions of ownership relate more easily to rain than to large-scale public diversions
Focus on blue surface water: water found in rivers, and lakes	Focus on green water: water stored in the soil profile or blue water stored in aquifers
Most water users get water from 'service providers;' most water provision is in the formal sector-making water resources governance feasible	Most water users get their water directly from rain and from private or community storage without any significant mediation from public agencies or organised service providers. Because the bulk of water provision takes place in the informal sector, it is difficult to pass enforceable water legislation
Small numbers of large-scale stakeholders	Vast numbers of small-scale stakeholders
Low transaction costs for monitoring water use and collecting water charges	High transaction costs for monitoring water use and collecting water charges

Source: Shah, Molden and Sakthivadivel (2003); http://www.iwmi.cgiar.org/home/integrated_river_basin.htm, accessed November 2004.

environmental management, to mediate conflict and build strong working relationships between disparate players.

Principle 6: Prioritising actions

Integrated river basin management will be more likely to succeed when short-term actions (say within 3 years) are implemented, visible results change the landscape and water quality improves. This requires clear identification of these actions and immediate commitment to action by river basin managers. These actions need to be specified within a River Basin Management Plan (see Chapter 4). It is unlikely that this Plan will be developed within a short time-period, so interim river basin management actions should be designed and implemented immediately. Long-term river basin management planning can be commenced simultaneously with a goal of developing an agreed, cost-shared plan of action in a 3 year time frame.

Principle 7: Accountability

A process of accountability is required to monitor the effectiveness of a river basin management plan and the organisation responsible for its implementation. This task can be implemented at the commencement of a river basin management plan and be linked to a river basin-based State of the Environment report. In this way, regular reports (say every 2 years) chart the progress of river basin health in, for example, critical water quality indicators. Similar organisational performance indicators can be developed and used to analyse the effectiveness of an RBO (see Chapter 7). In any jurisdictional context, one option is to undertake a gap analysis as the first task—to identify the gaps between policy development and implementation. This involves using various policy analysis tools to clarify these gaps and to develop statements of principles and courses of action to achieve gap reduction.

Principle 8: Local government partnerships for effective implementation

There is continuing concern about the role and ability of local government to implement local forms of river basin management. Local government has a key role to play in local governance—decisions which can have a more immediate impact on resource conditions. Planning and local zoning mechanisms are a valuable mechanism which can be used to implement broader river basin management goals (see also Chapter 4).

Local Government powers should be harnessed within a River Basin Management Plan to enact IRBM, and implement it through sub-basin plans. Such actions need to be congruent with the objectives of a River Basin Management Plan. Progress has been made in Ontario, Canada, by using Subwatershed Planning (Watershed Planning Implementation Project Management Committee 1997) to enact river basin management. This involves the use of a hierarchical planning process from Watershed Plans to Subwatershed Plans and Site Management Plans. The Resource Management Act (1991) of New Zealand provides specific functions for catchment management to the 16 regional authorities of that country. They can establish, implement and review objectives, policies and methods to achieve integrated management of the natural and physical resources their region. The Ontario and New Zealand experiences provide valuable models for regional and local natural resources management within an IRBM plan.

Principle 9: Integrating functions for coordinated river basin management

Lack of coordination between and within government agencies, NGOs, the general public and locally and regionally significant water stakeholders is a constant problem in IRBM. The solution lies in identifying integration, coordination and planning mechanisms and driving coordination throughout the RBO and with its strategic stakeholders. Coordination mechanisms are listed below, but the starting point is to establish a joint Vision for the basin and an ethic of willingness to cooperate, coordinate and manage together. This requires clear specification of the roles and responsibilities of partners in any joint action.

2.3.2 Functions

There are many functions of a river basin management organisation at both the international and national level, depending on the context of natural resources management and environmental planning. The following checklist provides an array of functions which can be used.

Regional natural resources management planning

This includes policies and plans for management of land and water resources across the river basin. The plans set out the strategic approach to natural resources management required in the river basin, including:

(a) an appraisal an evaluation of natural resources and their condition and trend;
(b) an analysis of community needs;
(c) subcatchment goals;
(d) subcatchment implementation guidelines;
(e) details of cost-sharing programs for on-ground works and other actions;
(f) details of a monitoring program and
(g) appendices which describe special catchment management issues, areas, management techniques.

River basin management protocols and plans are discussed further in Chapter 4.

Coordination mechanisms

RBOs have a fundamental role to coordinate decision-making about natural resources management. Table 2.5 provides a list of integration mechanism tools which can be used by river basin managers at different levels. They include planning, conflict resolution and communication tools.

Social assessment, social impact assessment and public involvement

An RBO requires data on the demography, social networks and human resources of the basin, which are best presented in a basin atlas. The data will provide information to assist basin management decision-making, data such as the capacity of resource managers to change, analysis of social change indicators, analysis of adoption rates by all resource users of best management options. Effective basin organisations will develop ex-ante and ex-post social impact assessment procedures for implementing basin management plans and will therefore require data sets to monitor the condition and trend of the social decision environment. There will also be data sets which describe the information management behaviour and delivery mechanisms for basin management plans (see also Chapters 5 and 6).

Table 2.5 Mechanisms for improving coordination in IRBM

Tools for joint planning and management	Tools for resolving conflict	Tools for communicating
Joint forecasting or scenarios Joint models or jointly used geographic information systems Co-location of personnel or creation of common jurisdictional boundaries Joint review of plans or environmental impact statements Formal review of clearance procedures Supervisory oversight Joint budgeting process Coordination committees Joint staffing or joint staff work groups Joint permit reviews or common standards for review Joint planning process (including environmental impact assessments) Cost-sharing arrangements for financing river basin management works Joint plans of action (projects, programs, policy, other)	Additional research or analysis Interpersonal or inter-group communication Appeal to higher authority Special meetings of committees or other groups Negotiation/bargaining within the group Appeal to outside party or third party (facilitation, mediation, etc.) Use of community advisory committees International water agreements Village level meetings and use of tribal customary law	Information and data sharing procedures Common database or data gathering Regular communication mechanisms (e.g. newsletters, e-mail) Scheduled meetings Intranet for joint development of plans, papers Informal communication, social occasions, word of mouth networks

Source: Adapted from Margerum and Born (2000), Margerum and Whitall (2004).

Natural resources inventory

The description of condition, trend and spatial location and variability of natural resources is a fundamental tool for IRBM. These data and the information and knowledge they provide will form a first step understanding of the conditions of natural resources, on which sound judgements can be made about prioritising natural resources management. The resource inventory can also be used to monitor the state of the environment through time in the river basin, so that changes in resource conditions resulting in river basin management plans can be identified. This will form a feedback loop to the decision process. These data sets can be developed in a geographic information system, ideally owned and operated by the RBO in partnership with government agencies, NGOs and other leading basin management stakeholders.

Legislative instruments and policy review

RBOs do not exist in an institutional vacuum. They exist in an often turbulent institutional environment. RBOs can have a powerful function in identifying and reviewing (or calling for the review of) current relevant legislation including local planning and natural resources management and environmental legislation, relevant to the river basin. This includes international agreements which impact on these two other forms of legislation, if it is an international RBO. This procedure may result in the identification of needed legislative reforms, and the need for new policies and high-level (perhaps Cabinet-level of government) initiatives between ministers of participating governments and departments. A policy instruments database is useful in cataloguing current legislation and policy instruments relevant to river basin management within and between countries.

Decision support infrastructure

The application of decision support systems, information management systems and models to evaluate resource management options is a fundamental task in river basin management. There is a wide range of such tools available to an RBO and they are discussed in detail in Chapter 5. There is the opportunity to use stakeholder partnerships to develop resource use scenarios, criteria and variables in models and decision support systems. These tools, if used appropriately, can enhance the knowledge level of the RBO and also be used as a mechanism to engage stakeholders and provide a mechanism for great ownership of the process of prioritising natural resources management goals.

Regional economy inventory

In many countries, RBOs have the opportunity and functionality to develop basin level indicators of economic performance. While they do not have to be the collector of such data, their added value role is to assemble these data, interpret the data and provide information about how proposed natural resources management goals will impact on the basin's economy. Such information should be stored in a regional atlas and/or a web-based geographic information system (see Chapter 6). The RBO's role is to facilitate the analysis of drivers of landuse change and water resources developments, and provide a scenario-building capability for the basin community and relevant national and international governments.

Information Management System Infrastructure

Allied to these functions there is the development of a basin-wide information system. The details of such a system are discussed in Chapter 6 and exist within a broader river basin information exchange process.

Knowing these principles and functions, it is then possible to derive best practices for good river basin governance, practices which can be applied to a wide range of settings. These are discussed in Chapter 4 (plans and protocols), Chapters 5 and 6 (information, modelling, etc.) and Chapter 7 (social, institutional and performance evaluation tools).

2.4 SUMMARY

In this chapter, we examined different meanings of the 'river basin'. Nine types of RBOs in river basin management were identified: advisory committee, authority, association, commission, council, corporation, tribunal, trust and federation. The functions of these different RBOs vary according to the context and need for institutional development. Generic attributes for effective IRBM were provided. These are built on a range of mechanisms and tools within the context of creating an enabling environment and building institutional capacity.

It is critical to build the capacity an RBO to be the governance leader in a river basin decision-making process. To do this, RBOs need the support of technical competence, an adaptive management style which addresses recurrent and priority problems, broad stakeholder involvement, sustained financing and clear jurisdictional boundaries and appropriate powers.

The focus of this chapter is to improve governance at the basin level, but what is *governance*? It is *a suite of decision processes in natural resources management within the basin and external to the basin which impact on the basin's natural resources management.* These are discussed in more detail in the next chapter.

2.5 REFERENCES

AACM International & Centre for Water Policy Research. 1995, Enhancing the Effectiveness of Catchment Management Planning. Final Report. Final Report for the Department of Primary Industries and Energy. Adelaide, AACM International Pvt. Ltd.

Barrow, C. J. 1998, River basin development planning and management: A critical review. World Development 26[1], 171–186.

Bellamy, J. A. & Dale, A. 2000, Evaluation of the Central Highlands Regional Resources Use Planning Project: A Synthesis of Findings. Brisbane, Australia, CSIRO Sustainable Ecosystems.

Bellamy, J. A., McDonald, G. T., Syme, G. J., & Butterworth, J. E. 1999, Evaluating integrated resource management. Society and Natural Resources 12, 337–353.

Bellamy, J. A., Walker, D. H., McDonald, G. T., & Syme, G. J. 2001, A systems approach to the evaluation of natural resource management initiatives. Journal of Environmental Management 63[4], 407–423.

Bent, M. 1998, Dawson River Catchment Strategy. Theodore, Queensland, Dawson Catchment Coordinating Association.

Blackmore, D. J. 1995, The Murray–Darling basin initiative: A case study in integrated catchment management. Sustainable Agriculture Resource Base. Office of the Chief Scientist, Department of the Prime Minister and Cabinet, Canberra.

Born, S. M. & Genskow, K. D. 1999, Exploring the "Watershed Approach"—Critical Dimensions of State-Local Partnerships: The Four Corners Watershed Innovators Initiative Final Report. Madison, WI, USA, Department of Urban and Regional Planning, University of Wisconsin—Madison.

Born, S. M. & Genskow, K. D. 2000a, The Watershed Approach: An Empirical Assessment of Innovation in Environmental Management. 7. Washington, DC, National Academy of Public Administration.

Born, S. M. & Genskow, K. D. 2000b, Towards Understanding New Watershed Initiatives: A Report from the Madison Watershed Workshop. 1–23. Madison, Wisconsin, University of Wisconsin-Madison Cooperative Extension Service.

Born, S. M. & Genskow, K. D. 2001, Towards Understanding New Watershed Initiatives: A Report from the Madison Watershed Workshop. 1–23. Madison, Wisconsin, University of Wisconsin-Madison Cooperative Extension Service.

Born, S. M. & Sonzogni, W. C. 1995, Integrated environmental management: Strengthening the conceptualization. Environmental Management 19[2], 167–181.

Burchi, S. 1985, Different types of river basin entities—A global outlook. Lundqvist, J., Lohm, U., & Falkenmark, M. (eds.). Strategies for River Basin Management. Dordrecht, D. Reidel.

Burton, J. R. 1984, The Art of Resource Management. Armidale, Resource Engineering Department, University of New England.

Burton, J. R. 1985, Development and Implementation of Total Catchment Management Policy in New South Wales. Sydney, Australia, Inter-departmental Committee on Total Catchment Management.

Burton, J. R. 1986, The total catchment concept and its application in New South Wales. Hydrology and water resources Symposium. Griffith University, Brisbane.

Burton, J. R. & Aust.F.I.E. 1988, Catchment management in Australia. Civil Engineering Transactions 30[4], 145–152.

Chave, P. 2001, The EU Water Framework Directive. An Introduction. London, IWA Publishing.

Cole, R. A., Feather, T. D., & Munch, J. D. 2005, Watershed planning and management in the United States. Bruins, R. J. F. & Heberling, M. T. (eds.). Economics and Ecological Risk Assessment: Applications to Watershed Management. 2nd Edition. 11–42. Boca Raton, FL., CRC Press.

Department of Water Affairs and Forestry. 1996, The Philosophy and Practice of Integrated Catchment Management: Implications for Water Resources Management in South Africa. TT 81/96. Pretoria, Water Research Commission.

Dieperink, C. 2000, Successful international co-operation in the Rhine catchment area. Water International 25[3], 347–355.

Dietz, T., Ostrom, E., & Stern, P. C. 2003, The Struggle to Govern the Commons. Science 302, 1907–1912.

Downs, P. W., Gregory, K. J., & Brookes, A. 1991, How integrated is river basin management? Environmental Management 15[33], 299–309.

Environmental Protection Agency. 1995, The Watershed Protection Approach 1993/1994 Activity Report. Washington, DC, USA, EPA.

Environmental Protection Authority. 1993, Geographic Targeting: Selected State Examples. Washington, DC, Environmental Protection Authority, USA.

Global Water Partnership. 2002, Integrated Water Resources Management ToolBox. Stockholm, Global Water Partnership.

Global Water Partnership Technical Advisory Committee. 2000, Integrated Water Resources Management. TEC Paper No. 4. Stockholm, Global Water Partnership.

Harden, G. 1968, The tragedy of the commons. Science 162, 1243–1248.

Healthy Rivers Commission of New South Wales. 2003, Healthy Rivers for Tomorrow. Sydney, Healthy Rivers Commission of New South Wales.

Heathcote, I. 1998, Integrated Watershed Management: Principles and Practice. New York, Wiley.

Hillman, M., Aplin, G., & Brierley, G. 2003, The importance of process in ecosystem management: Lessons learned from the lachlan catchment, New South Wales, Australia. Journal of Environmental Planning and Management 46[2], 219–237.

Hooper, B. P. 1995, Best Management Practice for Integrated Land and Water Management. Final Report on Project. #M334 to the Murray–Darling Basin Commission. Armidale, Australia, Centre for Water Policy Research, University of New England.

Hooper, B. P. 2001, Information management behavior of Australian catchment managers. Water International 26[4], 513–526.

Hooper, B. P., McDonald, G. T., & Mitchell, B. 1999, Facilitating integrated resource and environmental management: Australian and Canadian perspectives. Journal of Environmental Planning and Management 42[5], 746–766.

Imperial, M. T. 2004, Collaboration and performance management in network dettings: Lessons from three watershed governance efforts. Lawrence, P. & Kamensky, J. M. (eds.). Washington, DC, IBM Center for The Business of Government. Collaboration Series.

Jacobs, J. W. 1999, Comparing river basin development experiences in the Mississippi and the Mekong Water International. Water International 24[3], 196–203.

Jaspers, F. G. 2003, Institutional arrangements for integrated river basin management. Water Policy 5, 77–90.

Jonch-Clausen, T. & Fugl, J. 2001, Firming up the conceptual basis of integrated water resources management. International Journal of Water Resources Development 17[4], 501–510.

Kenney, D. S. 1997, Resource Management at the Watershed level: An Assessment of the Changing Role in the Emerging Era of Community-Based Watershed Management. Boulder, Colorado, Natural Resources Law Center, University of Colorado School of Law.

Lang, R. 1986a, Achieving integration in resource planning. Lang, R. (ed.). Integrated Approaches to Resource Planning and Management. 27–50. Calgary, Canada, The Banff Centre for Continuing Education.

Lang, R. 1986b, Integrated approaches to resource planning and management. Lang, R. (ed.). Calgary, Canada, Banff Centre for Continuing Education.

Le Moigne, G., Subramaniyan, A., Mei, X., & Giltner, S. 1994, A Guide to the Formulation of Water Resources Strategy. 263. Washington, DC, World Bank. World Bank Technical Paper.

Llamas, M. R. 2003, Lessons Learnt from the Impact of the Neglected Role of Groundwater in Spain's Water Policy. Water Resources Perspectives: Evaluation, Management and Policy. 63–81. Amsterdam, Elsevier Science.

Margerum, R. D. 1995, Integrated management for the Wisconsin Department of Natural Resources: An Overview. Wisconsin, Wisconsin Department of Natural Resources. Bureau of Water Resources.

Margerum, R. D. & Born, S. M. 2000, A coordination diagnostic for improving integrated environmental management. Journal of Environmental Planning and Management 43[1], 5–21.

Margerum, R. D. & Whitall, D. 2004, The challenges and implications of collaborative management on a river basin scale. Journal of Environmental Planning and Management 47[3], 407–427.

Millington, P. 1999, River Basin Management—Its Role in Major Water Infrastructure Projects (Draft). Cape Town, South Africa, World Commission on Dams Secretariat. World Commission on Dams Thematic Reviews. Institutional Processes V.3 River Basins—Institutional Frameworks and Management Options.

Millington, P. 2004, Integrated River Basin Management—From Concepts to Good Practice, Briefing Notes to the World Bank (Draft).

Mitchell, B. 1983, River basin planning: Problems, opportunities and implications for Ontario conservation authorities. Canadian Water Resources Journal 8[1], 1–20.

Mitchell, B. 1986, The evolution of integrated resource management. Lang, R. (ed.). Integrated Approaches to Resource Planning and Management. 13–26. Calgary, Canada, The Banff Centre for Continuing Education.

Mitchell, B. 1987, A Comprehensive-Integrated Approach for Water and Land Management. [No. 1]. Armidale, NSW. Australia, Centre for Water Policy Research, University of New England. Occasional Paper.

Mitchell, B. & Hollick, M. 1993, Integrated catchment management in Western Australia—Transition from concept to implementation. Environmental Management 17[6], 737–743.

Mostert, E., Van Beek, E., Bouman, N. W. M., Hey, E., Savenije, H. H. G., & Thissen, W. A. H. 1999, River Basin Management and Planning. Keynote paper for International Workshop on River Basin Management. The Hague, International Workshop on River Basin Management. International Workshop on River Basin Management.

Murray–Darling Basin Commission. 2001a, Integrated Catchment Management in the Murray–Darling Basin. 2001–2010. Delivering a Sustainable Future. Canberra, Murray–Darling Basin Ministerial Council.

Murray–Darling Basin Commission. 2001b, Murray–Darling Basin Initiative. Canberra, Murray–Darling Basin Commission.

Murray–Darling Basin Ministerial Council. 1987, Murray–Darling Basin Environmental Resources Study. Canberra, Murray–Darling Basin Commission.

Ostrom, E. 1990, Governing the Commons. The Evolution of Institutions for Collective Action. New York, Cambridge University Press.

Ostrom, V. 1971, An Alternative Approach to the Design of Public Organizational Arrangements. National Water Commission.

Radosevich, G. E. & Olson, D. C. 1999, Existing and Emerging Basin Arrangements in Asia: The Tarim Basin Water Resources Commission and the Mekong River Commission. Third Workshop on River Basin Development: TBWRC and MRC-World Bank, WDC, 24 June 1999. 1–7. Washington, DC, World Bank.

Rogers, P. 1993, The value of cooperation in resolving international river basin disputes. Natural Resources Forum 17[2], 117–131.

Rogers, P. & Hall, A. W. 2003, Effective Water Governance. Stockholm, Global Water Partnership.

Saha, S. K. 1979, River basin planning in the Damoday Valley. Geographical Review 69, 273–287.

Shah, T., Molden, D., & Sakthivadivel, R. 2003, Limits to Leapfrogging: Issues in Transposing Successful River Basin Institutions in the Developing World. Colombo, Sri Lanka, IWMI-TATA Water Policy Program. Water Policy Briefing.

Shrubsole, D., Hammond, V., & Green, M. 1994, Assessing the Implementation of Section 28 of the Conservation Authorities Act in London, Ontario (1978–1989). London, Ontario, Department of Geography, University of Western Ontario.

Siann, J. M. 1981, Conflicting interests in river basin planning. Saha, S. K. & Barrow, C. J. (eds.). River Basin Planning: Theory and Practice. 215–232. Chichester, UK, John Wiley and Sons.

Smith, C. T. 1969, The drainage basin as an historical basis for human activity. Chorley, R. J. (ed.). Water, Earth and Man. 101–110. London, Methuen and Co.

Sturgess, G. L. 1997, An Institutional Analysis of Catchment Management in the Hawkesbury Nepean River System. 1002. Sydney, Healthy Rivers Commission of New South Wales. Occasional Paper.

Synnott, M. 1991, A review of ICM in Australia. Land and Water Research News [8], 42–47.

Tardieu, H. 2004, Irrigation and Drainage Services. Some Principles and Issues Towards Sustainability. An ICID Position Paper. New Delhi, International Committee on Irrigation and Drainage (ICID). ICID.

Van Ittersum, M. & Van Steenbergen, F. 2003, Ideas for Local Action in Water Management. Stockholm, Global Water Partnership.

Watershed Planning Implementation Project Management Committee. 1997, An Evaluation of Watershed Management in Ontario. Toronto, Ontario, Ministry of Environment and Energy.

3

Governance in IRBM: a decision-making process

3.1 GOVERNANCE IS A DECISION PROCESS

3.1.1 The tragedy of river basin 'commons'

A river basin can be seen as a place where 'common-pool' natural resources management occurs. In a river basin, natural resources are frequently shared amongst multiple users. Let us take, for example, the 'common property' of the water stored in a reservoir which supplies a metropolitan city. Let us assume that the reservoir's catchment area includes private and publicly owned lands and that both land types are the sources of nutrients into the reservoir. Over many years, there has been increasing build up of nutrient levels in the waters of the reservoir. Whose problem is this? Is it the problem of the reservoir manager who provides water to the metropolitan city? Is it the problem of the recreation users of the reservoir who like to swim there every summer, but the waters are becoming eutrophic and present a health hazard? Is it the problem of the intensive feedlot producers in the catchment whose effluent has gradually flowed into the reservoir for many years? Is it the responsibility of the suburban residents who rely on a local council authority to construct facilities which remove stormwater, including high nutrient runoff from their gardens, into the local stream and then downstream to the reservoir?

 ISBN: 1 84339 088 4.

The answer, of course, is that it is everyone's problem, but no-one's problem. There is no specific organisation which is responsible for the management of quality of water flowing into the reservoir and which can influence the actions of a multitude of organisations and individuals who all play a role in contributing to the water quality problem. Thus, it is a 'tragedy of the commons' (Harden 1968).

In this chapter, we explore decision-making about a 'water common' (the river basin). Decision-making is the core to governance: it is the process by which resource managers at all levels and within different organisations make choices about the allocation and use of natural resources. The discussion about 'commons' management is further developed in the last chapter. But first, let us look at decision-making and its origins in the fields of geography and natural resources management.

3.1.2 Origins of decision theory: geography and natural resources management

The human–environment tradition of geography is a good place to start when discussing a disciplinary context for decision-making in river basin management. Geography has a rich tradition of decision-making studies. Geographers explored the ways people make choices about resource use, and how these choices reflect their perceptions of those resources. Geographical research was influenced by the work of environmental psychologists, ecologists and resource economists to form the foundation of a modern natural resource management paradigm upon which IRBM is built.

Decision-making—a process of choice selection amongst variables to achieve desired outcomes. The decision process is a conscious procedure, involving judgement, preference and commitment, whereby desired outputs are sought from a limited set of perceived resource combinations through the choice among various managerial, technical and administrative alternatives.

One of the first people to examine the character of natural resources and how decisions were made about their use was Zimmerman (1951) who claimed that "resources aren't, they become." He meant that resources were those parts of the natural environment perceived to be of use, to satisfy human needs and wants. From this perspective, the characteristics of natural resources change depending on human perceptions of them.

What then is natural resources management? It can be defined as the manipulation of resource-producing natural systems to optimise their long-term productivity for both human use and biological production. It is a *decision process*, best done in the overall context of sustainable landuse and which needs to be undertaken in an ecosystem context for ecologically sustainable development (Burton 1984). 'Wise' landuse means the use of natural resources to avoid degradation, that is, using land within the constraints imposed on landuse by the inherent biophysical characteristics of land itself. A comparable definition of resource management is a decision-making process. It involves allocating resources according to the needs, aspirations and desires of people within the framework of society's technological inventiveness, political and social institutions and legal and administrative arrangements (O'Riordan 1971).

Burton (1984) claimed that water resource management is best contextualised in the concept of a 'land–resource–environment' interaction system. The nature of resource management is bound up with how people use land and water resources to gain utility; the management of this interactive phenomenon is the concern of geography and natural resources management. Geographers focussing on people–environment relationships and natural hazards (floods, droughts, weather extremes) have provided new approaches to the practice of water resources management. Since the middle of the 20th century, the scholarly leadership in geography came from a focus on natural resources management and natural hazards—lead by scholars such as Gilbert White, Robert Kates and Ian Burton (Burton 1961; White 1961; Kates 1962; White 1963; Burton and Kates 1964; Burton, Kates and White 1968; White 1970; Kates 1971; White 1974; Burton, Kates and White 1978; Burton, Kates and White 1993; White 1997).

Geographers played an important role in natural resources management as a method of assessing resource potential and planning for resource use. Natural resources management shares a concern with this geographical tradition, as it focuses on people–environment interactions, but it does not necessarily focus on the spatial dimensions which geographers pursue.

Much work in resource management by geographers emerged with the behavioural revolution in that discipline in the 1960s. For example, White (1963) maintained that the study of resources was fundamental to the geographic tradition. In the context of resource use, he stated that:

> . . . what does seem important is to recognize the intellectual problems which call for solution and which because of their relation to spatial distributions and human adjustment to differences in the physical environment are of interest to geographers.
>
> (White 1963, p. 426)

White (1961) probed the meaning of choice of use in resource management. He developed a framework for describing resource decisions from research in floodplain occupance, water use and recreation landuse. He suggested that where resource managers made a choice about a specified resource, they evaluate some or all of the quantity and quality of the resource; the present values of the gains and losses accruing from future use of the resource; the technological change which might affect future demand, production and compatible uses and the relation of the resource to other resource uses in contiguous or functionally linked areas.

White emphasised that

> perception of environments is a basic feature of resource management and may drastically limit the practical range of choice. From this starting point through appraisal of possible uses, income streams, technological trends, and regional impacts, the comparison of the manager's appraisal with that of others helps identify distinctive and crucial aspects of decision-making
>
> (White 1961, p. 29)

He showed how decisions are made, not from a theoretical range of choice set by the physical environment, but from the practical range of choice set by culture and institutions. Other geographers, for example, Golledge, Brown and Williamson (1972), noted that the behavioural approach developed two significant streams of thought in the 1960s. The first was an emphasis on 'man–land' environment relations expressed through human perceptions of environment. The second approach was more targeted, focussing on the goals, aspirations and motivations of decision-makers. Wolpert (1964) examined the decision process in a spatial context. He suggested that decisions were made not following an optimising procedure, but rather under sub-optimal conditions of imperfect knowledge. This approach was stimulated by the concept of satisficing rather than optimising decision-making behaviour (Simon 1957). Wolpert and Simon's work mark a watershed in thinking in decision-making research. Their questioning of the accepted dominant view of the time, that decision-making was concerned with 'Economic Man' making 'rational' decisions based on perfect knowledge, led to a significant reformation of conceptual frameworks for decision-making in natural resources management. The emphasis was now placed on a broader range of variables that influence the decision process.

While natural resources management may be an ally of geography, it has evolved as a divergent field of enquiry, drawing techniques from many fields. Johnson (1985) explored the links between natural resources management, human and physical geography and resource analysis, but commented that these links were yet to be proven. Geography has tended to draw from other fields of enquiry—hydrology, water resources engineering and water resources economics—and provided a spatial perspective on the management of natural resources. Much of the early effort of collaboration was originally focussed on the assessment of water resources potential for river basin development—including the estimation of physical potential, the determination of technical and economic feasibility and the evaluation of social desirability (Chorley 1969).

3.1.3 Changing approaches to decision-making

There have been substantial changes to approaches to water resources decision-making in recent years, with the emergence of new paradigms, as outlined in Chapter 1. Traditional approaches were essentially hydro-centric. They were single sector (water) oriented in which the river basin or groundwater province was viewed as a complex physical system—based on complex interrelationships between the hydrological and geomorphologic characteristics of the basin and its rivers and streams. This approach, common in the 1930s to 1960s and favoured by water engineers and water economists, viewed the basin as a water resources system whose water resources were to be exploited for economic development. Decision-making emphasised the determination of maximum possible yield and developing mechanisms for most effective water allocation between users. It was used for significant water resources development projects, such as the Hoover Dam project in the United States—an era characterised by dam building and irrigation expansion in very large water resources projects. The single sector approach was driven by highly scientific methods and technological innovation, with an overall purpose of maximising available yield from river basins and watersheds. More complex decision approaches promoted multi-objective development of water resources systems including recreation, hydropower, navigation and irrigation development, as evidence in the work of the Tennessee Valley Authority and the US Army Corps of Engineers in the United States, the Nagarjuna Sagar Dam project in India and the Snowy Mountains Scheme in Australia.

This approach was questioned in the 1970s, a period of rapidly rising environmental awareness and action. A new focus on ecosystems and the new science of ecology questioned the single or multi-objective approach to water resources management, with its strong development emphasis. The reality was that the traditional paradigm ignored the more diverse range of resource use features of river basins which interact to create the so-called 'wicked' problems of environmental management and sustainable water resources management. The new decision paradigm, integrated water resources management, recognised river basins as large, complex, integrated ecological systems. Many of the problems were called 'wicked' because they were intractable, beyond immediate solutions with currently available technologies and institutions.

The term 'ecosystem approach' was used as a corollary for the integrated approach. Here, the watershed was seen as an integrated ecological system in which human impacts are one component of the comprehensive functioning of ecosystems within a watershed. The geographer, Mitchell (1991), recognised that the challenge of this integrated approach was how such an approach was to be interpreted. He maintained that for too long advocates of an ecosystem approach had interpreted it to be synonymous with a *comprehensive approach*, in which attention is given to all components and linkages in a system. When a comprehensive approach is taken, the probability is very high that the period of time required to complete an analysis or a plan will be very long, resulting in the final plan often being no more than a historical document, because too many events or processes will have changed and made the plan obsolete before it is even completed.

Mitchell's interpretation of an *integrated approach* involves a more selective or focussed perspective. Not all components and connections in a system are considered, but only those which, on the basis of knowledge from all stakeholders (through focus groups or other forums involving people ranging from technical analysts to long-term residents) are judged to be the key drivers of variability in the system (Hooper, McDonald and Mitchell 1999). Both a comprehensive and an integrated interpretation are consistent with an ecosystem approach, but the latter leads to a more focussed approach and therefore increases the likelihood of a more practical output.

River basin stakeholder—a person or organisation with an interest in or affected by a river basin management decision.

3.1.4 Decision-making in river basin management

The nature of hydrological linkages suggests a river basin forms a natural unit for decision-making, especially in sub-humid, temperate, tropical and equatorial hydrological regimes where runoff occurs. Rivers are significant areas within watersheds, intimately linked to the land systems that surround them in a river valley. Rivers act as hydrological conduits receiving excess water from precipitation, infiltration and groundwater movement, and transfer water across the landscape to watershed outlets, such as another river, lakes, estuaries or oceans. The ecological health of a river system reflects the ecological health of the land systems in the river basin, indicating the impacts of upstream land management practices on ecological processes. Therefore, basin decisions on this areal dimension makes sense.

A recent statement on river basin governance captures the importance of the basin as a decision entity. The expert group statement on Integrated River Basin Management for the 2nd World Water Forum and Ministerial Conference in the Hague, 2000, maintained,

> water is an environmental resource and it is the basis for social and economic development. River basins are the paramount source of freshwater. To preserve and maintain this precious resource for present and future generations, there is the need for sustainable river basin management.

3.2 LEVELS OF DECISION-MAKING IN IRBM

3.2.1 A hierarchy of decisions

As discussed in Chapter 1, three levels of decision-making can be recognised in river basin management: individual, operational and strategic/policy levels. In a river basin dominated by agricultural operations, for example, representatives of the three levels would be farmers (individual), government agency resource managers (operational) and the policy makers in government (strategic/policy). In an urban setting, the three levels would represented by homeowners or industries (individual), environmental regulator agencies, city councils (operational) and state agency water pollution control agencies which set policy (strategic).

All three decisions-makers' functions vary in time and space:

- They vary according to the prevailing resource uses and economic environment which facilitates resource use development.
- They reflect the extent of institutional development of river basin organisations and river basin institutional arrangements (laws, policies, incentives, regulations).

- They reflect current government policy—in both roles and responsibilities.
- They change location according to changing functions of administration at different times.

Gregg et al. (1991) recognised this same three level hierarchy of decision-making in water resources management in an analysis of western USA water resources management. They recognised that where there were situations involving the use of natural resources, decisions tend to relate to the management of resource-based inputs and outputs in the production sectors of economies. These decisions are made by individuals, firms and other private and public sector organisations. They are driven by market signals, guidelines for resource use, political processes and operate within current institutional arrangements and administration for resource exploitation and conservation. Table 3.1 characterises the three levels and Figure 3.1 demonstrates the relationships between these levels.

Table 3.1 Characterising decision-making levels in river basin management

Level of decision making	Characteristics	Mapping scales for application	Examples
Policy—high level strategy development	Specifies the content of policy, means of implementation, means of changing policy Highest level institutions which set broad directions for national-scale natural resources management, at the basin level and for inter-basin and international river basin management agreements	Approximately 1:1,000,000+	Mekong River Agreement
Implementation—the organisational level of decision-making	Rules which constrain or enable decision-making at the operational level Second level institutions which specify the operationalisation of national and river basin scale policy and management guidelines May be composed of more than one level	Approximately 1:100,000	Murray–Darling Basin Natural Resources Management Strategy
Operational—the individual decision-maker	Shaped by the rules established at the implementation level On the ground actions	Approximately 1:25,000 to 1:10,000	Sub-watershed management plan

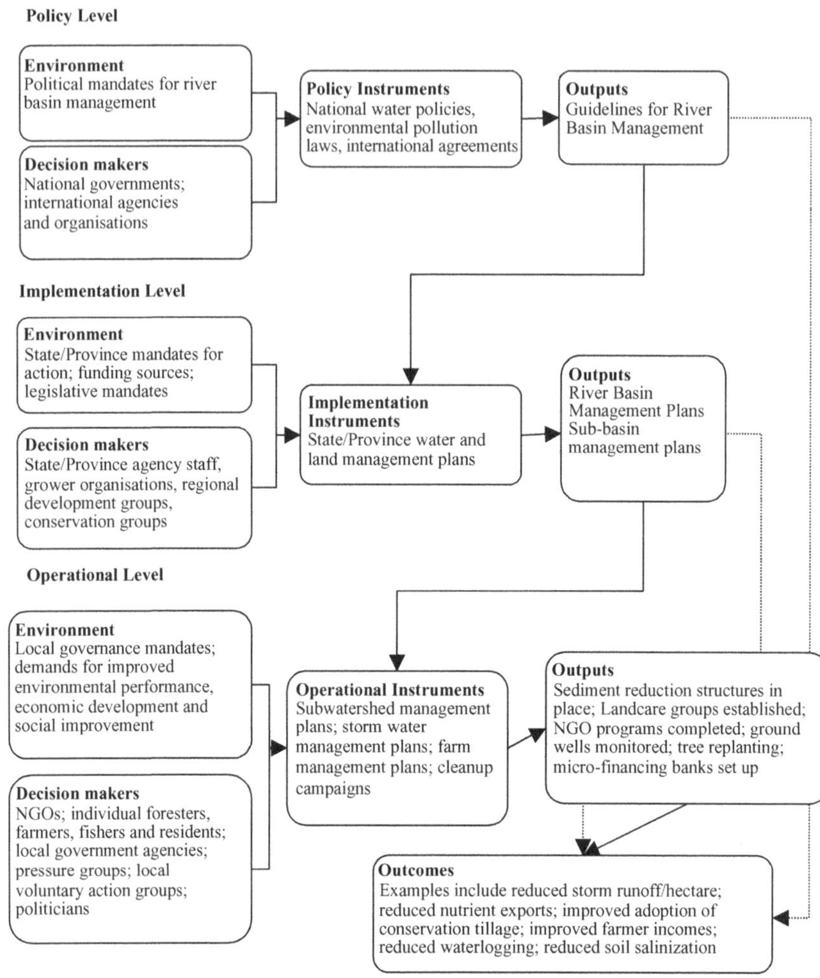

Figure 3.1 Decision-making levels in river basin management. Adapted from Gregg et al. (1991).

3.2.2 Case study: decision-making hierarchies in the Liverpool Plains Catchment, Australia

This hierarchical approach can be represented as a matrix of decision-making for river basin management. Table 3.2 shows a typical decision matrix using an Australian catchment within the Murray–Darling Basin. The Liverpool Plains Catchment is approximately 1200 km^2 in size (Figure 3.2). While not a river basin, the smaller size illustrates, even at this scale, that there are a large number and broad range of decision-makers in a sub-basin decision settings. When scaled

Table 3.2 Decision-making matrix for the Liverpool Plains Catchment, Namoi Valley, NSW, Australia

Scale of resource management	Private	Public
Local (operational)	*Farmers and graziers* Approximately 1500 farms *Rural businesses* —Several farmers operate off-farm business ventures —Suppliers and extension services (providers of agrochemicals, farm machinery, irrigation equipment and fertilizers, include consultants and advisers) —Transport (private stock and grain transport companies) *Landcare groups* —Supported by government grants, mainly local farmer-owned and organised	*Agricultural extension agents* —Extension services from resource management agencies, primarily Departments of Agriculture, Infrastructure, Planning and Natural Resources, (limited, and decreasing) —Private consultants (provide independent agronomic and on-farm financial advice; based within the region) *Shire officials* —Three Shires influence land ownership transfers, collect land taxes, local environmental management plans (Gunnedah, Quirindi and three properties in Murrurundi Shire)
Regional/ State (implementation)	*Businesses* —Banks (includes agricultural development banks, loan services) —Wholesalers —Services (providers of agrochemicals and fertilizers) —Transport (private stock and grain transport companies) *Agricultural extension and technical officers* —Chemical companies —Private consultants *Organisations* —Regional Development Boards —Private grower organisations (NSW Farmers, Grains Council of Australia) *Media*—Local and regional newspapers, television and radio stations (profile major resource management issues; influence attitudinal change; market products and services)	*State officials (agricultural extension and technical officers)* —Includes Department of Infrastructure, Planning and Natural Resources, Department of Agriculture, Environment Protection Authority, National Parks and Wildlife Service, State Rail, State Forests, Rural Lands Protection Board —Includes some regional policy and planning by government *Regional catchment management organisations* —Namoi Catchment Management Authority —Liverpool Plains Land Management Committee *Academics* —Social, economic and biophysical research scientists from local and regional universities and research field stations

Continued

Table 3.2 Decision-making matrix for the Liverpool Plains Catchment, Namoi Valley, NSW, Australia (*Continued*)

National (strategic/ policy)	*National businesses* —Banks (national policy affects borrowing capability, interest rates) —Wholesalers (impacts on product values and input costs) —Services (provision of consultancy skills) —Transport (provision of national infrastructure) *Organisations* —Private grower organisations (NSW Farmers, Grains Council of Australia) —Organic farming organisations *Media* —National newspapers, television and radio stations (profile major resource management issues; influence attitudinal change; market products and services)	*Philanthropic organisations* —Australian Conservation Foundation —Inland Rivers Network *Officials and programmes in federal organisations* —Murray–Darling Basin Commission —Land and Water Australia (an R&D organisation) —Rural Industries Research and Development Corporation —National Landcare Programme —National Dryland Salinity Management Programme *Media* —As for private (includes ABC TV and Radio)
Global (strategic/ policy)	*International agribusinesses* —None thought to be influential, although much agricultural produce is exported through national organisations to international markets. —Global market changes influence local farming practices (e.g. planting decisions)	*Academics and researchers* —None thought to be influential, although several international researchers use the Liverpool Plains as a comparative field site for research *Treaties* —Federal government requires compliance from states and regions with national policies derived from international agreements such as GATT and APEC, and global environmental initiatives including Ecologically Sustainable Development treaties *Philanthropic organisations* —None thought to be influential

Source: Hooper (1995).

up to the river basin of the Murray–Darling (1.1 million square kilometres), the number of decision makers increases at a very large order of magnitude.

There is a range of resource management issues in the Liverpool Plains Catchment, the most significant being the retention of native vegetation, floodplain management, soil salinity and improving water quality (Hooper 1995):

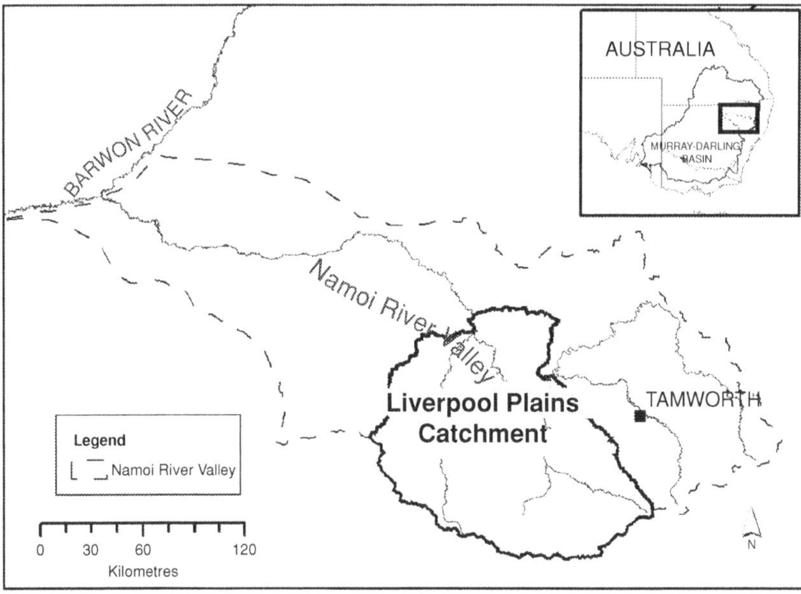

Figure 3.2 Location of Liverpool Plains Catchment in the Murray–Darling Basin.
Source: Murray–Darling Basin Commission

Only recent development of salinity management options

The development and implementation of salinity management practices on the Liverpool Plains are recent phenomena. Salinity management is a Landcare[1] initiative, with a significant planning focus coming from the Liverpool Plains Land Management Committee (the local sub-basin organisation). Technical issues (hydrogeological processes, salinity identification and monitoring, preferred agronomic practices and the on-farm economic analysis of salinity management options) are being analysed by various government departments. It is not fully known what links exist between salinity problems and increased shallow flooding that is occurring on the plains. Only very recent research has shown how salinity management relates to the changed surficial and soil water profiles that reflect changed climatic conditions (increased summer runoff) and changed land management practices.

Salinity management is based on generic land management units

Salinity management was developed as a set of best management practices based on the need to change land management practices, encouraging farmers to change from cropping production into pasture production, mixed crop and livestock production systems and opportunity cropping to maximise the use of water. There were eight generic land types in the catchment. Figures 3.3 and 3.4 illustrate two units. There are costs and benefits of alternative practices to manage salinity developed for each.

[1] Landcare is a process whereby farmer groups undertake local catchment management actions.

Figure 3.3 A portion of the Liverpool Ranges land management unit in the upper part of the Liverpool Plains Catchment.

These plateaux, ridges and slopes, occurring in the south and south east of the Liverpool Plains, are formed on tertiary basalt flows and have an average altitude of 1100 m. They make up the Liverpool Ranges. Median rainfall is around 800 mm. A marked vegetation change occurs below the plateau scarp and on the upper mountain and hill slopes. Tall open forests change to woodlands and to grasslands as altitude decreases; soils are shallow, medium to heavy clays. Landuse also changes with altitude from heavily wooded areas with minor hardwood logging in upper reaches, to extensively cleared areas for native and improved pasture grazing on hill slopes, and cultivation in grassland areas. Eight percent of these lands across the Liverpool Plains catchment are subject to high water tables (about half of the colluvial fans), but salinity is not a significant problem. *Photo source*: Author.

Figure 3.4 (a) and (b) Part of the low-flow alluvium land management unit of the Liverpool Plains Catchment.

Occupying 22% of the catchment, the land description is similar to high-flow alluvial plains but the hydrogeology is different. The low-flow alluvial areas are distinguished from the high-flow areas according to potential for irrigation, using flow rates for the Gunnedah aquifer. Flows <50 l/s are not suitable for irrigation. The low-flow alluvial areas are prone to waterlogging and salinity—62% of this land is subject to water tables less than 5 m below the surface. These are areas of concern for salinisation. *Photo source*: Author.

Implementation reflects an innovative farming culture

A process of landuse change is occurring with minimal formalised salinity management planning already in place. This reflects the innovative, dynamic approach to farming by the farmers of the Liverpool Plains. The process of moving to a pasture-based, opportunity cropping economy may reflect the increased opportunities and greater returns from beef production and maximising soil-water in the region.

Landuse changes reflect broader forces at work, not environmental hazards

Landuse changes on the Liverpool Plains were the result of economic forces causing farmers to move away from purely cropping enterprises into mixed farming or livestock production systems. This process is being driven by market forces, particularly the local (feedlot requirements), regional and national market demand for beef cattle and a diversity of opportunity crops. Salinity management may be an unexpected side benefit of a broader landuse change process. The landowners and Landcare group members have adopted local salinity management practices (tree planting, fencing off saline scalds), whereas market forces may be driving them into other practices that could have a salinity management benefit.

Landowners on the Plains are concerned about salinity and other environmental problems (as evidenced in the rapid rise of Landcare membership and attendance at floodplain management and water resources management public meetings in the 1990s). It is fortuitous that their positive environmental attitudes are congruent with opportunities to diversify into different farming enterprises which may have the side benefit of reducing the impact of dryland salinity.

High-quality but limited Landcare activities and extension programmes

The region has excellent extension agronomists and soil conservationists who form the core of an extension programme in salinity management. Extension and local leadership in salinity appear to be only limited by government funding. There are widespread enthusiasm and skills amongst resource management professionals.

These issues form the local context by which the decision-makers shown in Table 3.2 operate. The three levels of decision-making discussed in Table 3.2 form a model of the action arena for IRBM at the local scale, and are similar to the action arena of water resources management suggested by Gregg et al. (1991) for most water resources management contexts. The concern is with each of these levels: their specification, operation and relationships, and most importantly how they can be coordinated to improve integration.

In the next section, the focus is on the *individual (operational) and implementation levels* of decision-making. This involves understanding how individuals and organisations make decisions formed by the perception of the environment in which they operate and then discusses how a general model of decision-making in river basin management can be devised, using operational, implementation and policy levels.

This policy level is taken as a given in any situation, meaning that it sets the framework for decisions, often at the two lower levels in a decision hierarchy. In robust democracies, there are growing influences from political processes at all levels, and from individuals and organisations which tend to drive a bottom–up influence into the policy level. It is this mix of decision-making influences which makes integrated river basin management a challenge. Therefore, an understanding of the ways in which individual resource managers and organisations make decisions helps to clarify and to inform river basin management. In any decision setting, the following dimensions must be considered:

- The legal framework of the decision setting, which constrain actions;
- The biophysical characteristics of the catchment which influence resource use options—land characteristics determine opportunities for resource development and conservation and the capability of land determines the extent of landuse options beyond which land degradation occurs;
- Cultural features of the decision setting, which include the landuse traditions of the people;
- Individual psychological characteristics of the decision-makers;
- Administrative arrangements for natural resources management;
- The economic setting: current landuse practices which generate land and water management impacts and influence future options for economic growth;
- Financial constraints and incentives which provide limitations and opportunities for resource development and conservation.

3.3 CHARACTERISING INDIVIDUAL DECISION-MAKING

3.3.1 A choice process

The pioneering work of O'Riordan and Found characterised decision-making in natural resources management. Their work focussed on the role of the individual decision-maker and has been extended to examine the role of agency decisions. Found (1971) pointed out that perception, learning and decision-making were highly interrelated processes. He maintained that perception is important in the decision-making process as it influences learning processes which in turn determine the images of both the resources being used and the environment in which decisions are made. Learning in a landuse setting occurred as a response to two types of information: the individual knowledge gained by experience, and the knowledge from other sources through communication.

Applied to a river basin, these learning sources include personal knowledge and experience of past resource uses and market signals, institutional information on river basin management, community knowledge of natural resources management, and peer leader information. These knowledge sources are used

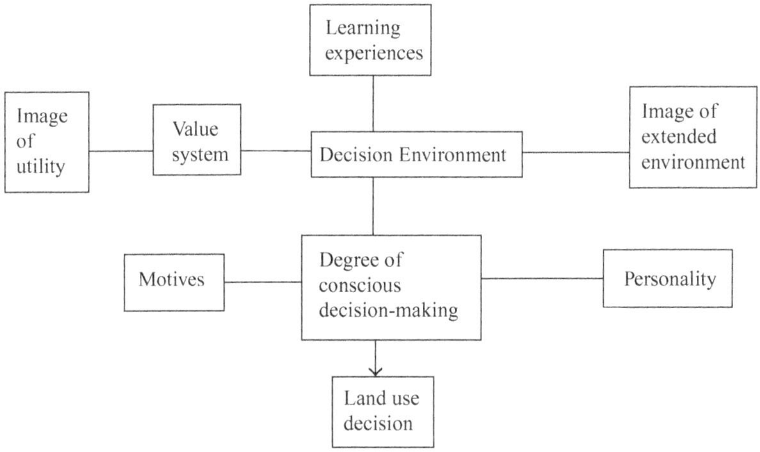

Figure 3.5 A simplified view of an individual's general decision framework. Adapted from Found (1971).

by resource managers to create images of the management of natural resources in a river basin, and produce landuse decisions in response to these images. The role and types of information are discussed further in Chapter 5.

Found developed a model that demonstrated how traditional economic and behavioural models could be viewed simultaneously, but indicated that they should be applied carefully according to the context of resource use (Figure 3.5). He suggested that resource use decisions working in an operational environment of pure competition and in a free enterprise economy may best be explained by rigid economic models. However in a different socio-economic context, say a subsistence agricultural economy, behavioural models may be more helpful in explaining decision-making.

A second early investigator of decision-making, O'Riordan (1971), defined decision-making as a means to allocate resources according to the needs, aspirations and desires of people within the framework of society's technological inventiveness, political and social institutions, and legal and administrative arrangements. This decision process is a conscious procedure, involving judgment, preference and commitment, whereby desired outputs are sought from a limited set of perceived resource combinations through the choice among various managerial, technical and administrative alternatives.

O'Riordan (1971) proposed a model to explain the decision-making process in natural resources management (Figure 3.6). The model was based on the work of a third pioneer, Gilbert White (1961) and the derivative studies of resource use in hazardous environments (listed in Section 3.1.2). The O'Riordan model demonstrated how decision-making is a learning process, with four principal stages in the process which are connected by feedback loops. Goals are formulated, resources are identified (through a number of signals), alternative strategies are formulated, then a choice is made depending on the evaluation of the perceived goals. The choice is finally evaluated.

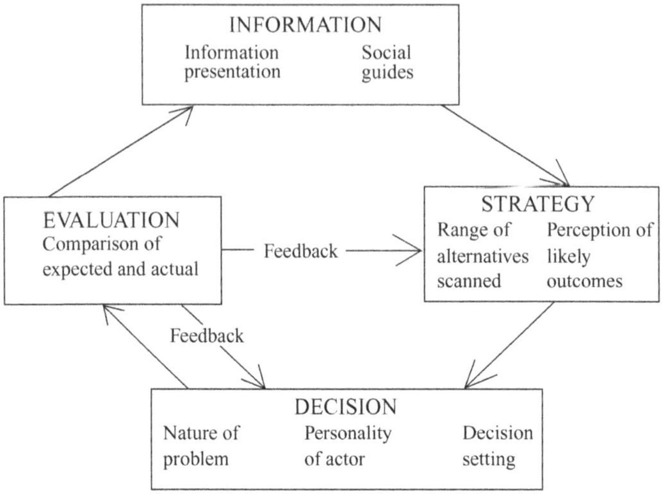

Figure 3.6 Model of natural resource management decision-making. Adapted from O'Riordan (1971).

Various institutional, social and personal factors are effective in the decision process (Figure 3.7). They were the cultural setting of resource use, the technological level available for use, the nature of the problem (how complex was the resource use), the previous experience and the personality of the resource manager. The model recognised the variety of behavioural, social and institutional factors influencing the decision-making process. Similar to Wolpert's 'satisficer' model (Wolpert 1964), O'Riordan demonstrated that people make decisions about using resources on more than purely economic or 'bounded rationality' terms.

3.3.2 The role of perception and cognition—a short history of research

Perception of environment is defined at the individual level as the ways in which a person or organisation receives signals from the environment through a number of senses—sight, touch, hearing, etc. People and at a broader scale, organisations, however, are selective in their choice of information received from these signals. Most are selective, only taking in a small part of the available information. Once information is received, it is organised, so that it 'fits in' with other information the person or organisation knows about the environment. This process is called *cognition*. The end product of perception and cognition is the *image* of whatever aspect of the environment the signals came from (Haynes 1980).

Research in environmental perception is not a new paradigm, but emerged with the behavioural revolution in geography described above. Saarinen (1966, 1969) examined the man–environment tradition in geography and discussed the role of perception in resource management and natural hazards research. He maintained that environmental perception was a new research frontier in geography in the

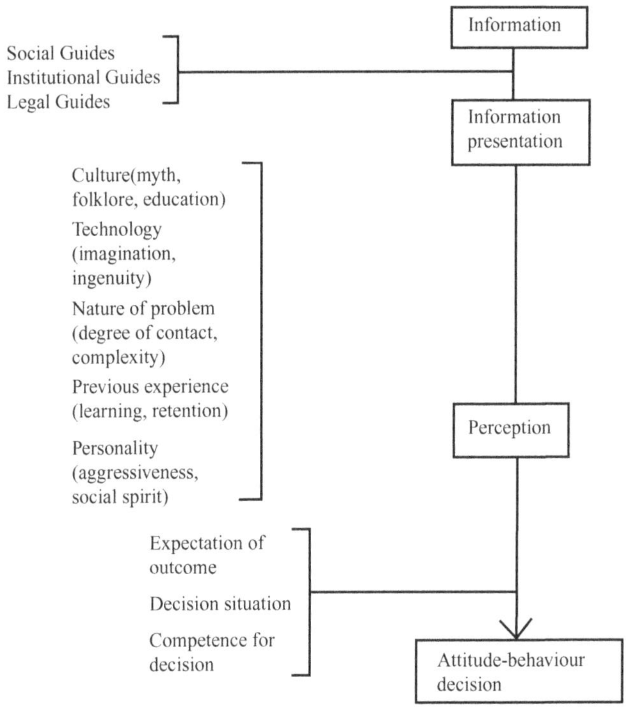

Figure 3.7 Factors affecting attitude and behaviour in the process of resource management appraisal.
Adapted from O'Riordan (1971).

1960s, drawing techniques from other behavioural sciences, but being used by geographers to understand the spatial dimension of resource use. He reviewed developments in perception research and defined the dimensions of both the perception and use of natural resources, and observed that:

- There are differences between optimal and actual resource use that were influenced by perception of resource managers;
- Cultural background affects the perceptions of resource managers;
- Perceptions of environmental gradients affected adjustment to resource use through different landuse practices.

Saarinen (op. cit) drew two conclusions. First, that there was a significant impact on individuals' decision environment by the influences of the impersonal environment. This included the images people formed of the natural world. Secondly, he noted that the way people perceived the natural world, how their perceptions were affected by experience, and how perceptions affect behaviour, all formed components of a more precise understanding of decision-making behaviour.

Mitchell (1971) developed a framework to relate environmental behavioural investigations at an individual level to macro-level water management issues, especially issues affecting the decision-making process of water managers. This was done using a case study of water pollution and decision-making in Ontario. Mitchell proposed that investigating the interaction between public perceptions of water pollution issues and the decision-making processes of those agencies responsible for managing water quality would aid the final resource allocation process. Mitchell's work was important at this stage in being one of the first studies to propose a method to incorporate public perceptions into decision-making processes in resource management.

Studies of resource use depend on an accurate understanding of the nature of the resource base in use. An ecological approach to landuse allows a more rational use of land, as the biophysical constraints inherent in land attributes play an important role in allowing or modifying landuse practices on them. However, it has also been shown from the empirical studies in environmental perception described above that this type of research further explains resource use. These findings are well known and stem to seminal research in the water sector for over 30 years. Pigram (1972) maintained, for example, that the changing perception of resources was the catalyst in introducing new irrigation practices in Australia. Perceptions of biophysical constraints changed with the introduction of new technologies and allowed a rapid reappraisal of resource use opportunities. These fundamental learnings of some 30 years ago are frequently lost in today's river basin management. It is vital to good basin governance to know the perceptions held by resources managers of resource constraints and opportunities.

Resource perception, including hazards perception, is the focus of interdisciplinary investigations in natural resources management, with an increasing research influence from environmental psychologists and social anthropologists. Saarinen, Sell and Husband (1982) listed the range of concerns of environmental perception to include planning and environmental problems, struggles to develop methodology beyond that encapsulated in the early North American hazard studies, and an emerging, unifying theme, the sense of place. This theme deals with,

> . . . what makes an environment . . . distinctive and different from others. It requires observation and analysis, both of the environments' features (both literal and symbolic) and of the perceiver's experience, reactions and values . . . The importance of the sense of place is its close tie to the sense of identity.
>
> (Saarinen, Sell and Husband 1982, p. 525)

The concept of sense of place in river basin management is useful because it asserts that the psycho-sociological characteristics of resource managers and the characteristics of the resources being used should be identified, observed and analysed to give a thorough understanding of decision-making.

More recently, a maturing environmental perception research sub-field was noted by Aitken (1991). He reported that perceptual and behavioural geography had 'come of age'. He provided evidence for this from the perplexing breadth of the research field and classified it into four sub-fields:

- spatial cognition and human behaviour;
- the ecological dimensions of resource manager–environment relations;
- landscape perception and experiences;
- comparative research involving varied social and cultural groups.

Table 3.3 Application of environmental perception and cognition research to river basin management decision-making

Sub-field	Application
Spatial cognition and human behaviour	Resource managers' knowledge of the size and location of their river basin Distance from main decision nodes (river basin organisations headquarters, agency stations) may influence willingness to participate in basin wide decision-making
The ecological dimensions of person–environment relations	Knowledge of impacts on ecosystem functions by resource management activities Use of this knowledge to change/modify resource management actions Incorporation of resource manager understanding of ecosystem functioning in decision-making processes at the local scale in sub-watershed management plans, river basin management plans, national water and land policies
Landscape perception and experiences	Different perceptions of landscapes exist between resource managers and between and within resource management and resource development agencies Varying perceptions generate resource management conflicts and call for an external process by the river basin organisation, to arbitrate the conflicts or develop agreed resource management solutions at the basin scale
Comparative research involving varied social and cultural groups	River basin management plans need to recognise the different social and cultural groups which exist in a river basin Varying perceptions of the efficacy of resource management methods (e.g. indigenous versus modern, western) exist within river basins and processes are required to incorporate these varying perceptions into agreed resource management plans

There has now developed a maturing field of geographic and water resource management enquiry based on environmental perception research. While it may be fragmented, there are several, at least these four sub-fields, to examine resource management behaviour in river basins (Table 3.3).

The river basin can be seen as a multi-resource use/multi-hazard, and a multi-scaled environment—a complex multi-faceted locus of natural resources management behaviour. Decision-making at the basin scale is influenced by the physical availability of resources, constraints on resource use, resource perceptions, the influence of natural hazards, agency arrangements, the roles of the private sector and non-government organisations and other characteristics of the biophysical, social and economic environment.

In this section, we examined conceptual models for understanding decision-making in river basin management. Later in this book, we examine computer-based models, some originating from the following conceptual models, to simulate and assist decision-making in river basin management. Here, the discussion is about the behaviour of individuals and organisations.

3.3.3 Models of individual decision-making and their application to river basin management

Three groups of decision models in natural resources management can be applied to river basin management:

(a) *People–environment interaction models*, which explain how resource managers make choices about location and resource use. These models vary and include explanations of how decisions about resource use are made in the context of finite knowledge, imperfect perceptions, and a complex array of goals—decision-making is a satisficing process. Others outline how multi-attribute preference models and associated measurement and estimation procedures can explain decision-making. Lynne, Schonkwiler and Rola (1988) showed how income level had a moderating effect on soil conservation behaviour, using the Fishbein–Ajzen model of attitude–behaviour relationships. Aitken (1991) maintained that attitude theory provides a basis for understanding decision-making: attitudes may indicate subsequent behaviour, but the decision to perform a specific behaviour is affected by a range of intervening variables (structural, societal and institutional constraints) which constrain the decision process.

(b) *Models of perception, resource use relationships and adjustments to natural hazards*, which focus on how resource managers' perceptions influence choice—the focus is on the role perceptions of natural hazards play in decision-making. The approach became known as the 'Chicago school' of hazard research, from were it emanated. The models demonstrated how individual resource management decisions are affected by the 'prison of experience' and led to many derivative studies, applying hazard research techniques to planning throughout river basins, such as Kates (1963). In these models, the emphasis is to explain how people adjust to natural hazards—decision-making regarding hazards was understood as an interactive, incremental adjustment process. Slovic, Kunreuther and White (1974) criticised the model because of its emphasis on rational decision-making and the tendency to overlook how people incorrectly perceive risks and replaced known probabilities with intuitively derived laws of chance. However, the learnings from the early models remain: that the perception of events, individual and community experience and the personality of resource managers act as factors which influence hazard adjustment strategies.

(c) *Models which analyse decision processes.* These are perhaps the most advanced of the three models; they attempt to explain decision-making processes relevant to river basin management. The models consider knowledge, attitudes and anticipations of the decision-makers as well as a range of economic variables and institutional requirements which affect decision-making. They recognise decision objectives and goals, the degree of rationality in the decision process, and the impacts of stress and threat in the decision-making environment as the focus of decision-making. The value of these models is that they critique the 'economic rationalist' view of decision-making as being unable to fully explain natural resources management behaviour. Here, attitudes are recognised as important to the decision process, and that structural, social and institutional constraints play roles as important than those of utility maximisation and volitional control.

The three models are valuable to river basin governance. They explain individual level of decision-making, rather than the strategic or operational levels as discussed earlier in this book. Three benefits result:

- It can help river basin managers to understand the reasons why individual resource managers (farmers, fishers, foresters, industrialists, urban dwellers) act in ways which support or work against broader scale river basin management practices.
- It provides understanding of how higher levels of decision-making can engage with the lowest level of governance.
- It illustrates the wide range of aspirations, attitudes, perceptions and social norms in society which influence individual behaviour. There is often complexity in understanding and experience in natural resources management, and this is often neglected when dealing with the general public.

3.4 THE ROLE OF PSYCHO-SOCIAL VARIABLES IN INDIVIDUAL DECISION-MAKING

There is growing acceptance of the need to devolve decision-making to the lowest possible level. Called, subsidiarity, this approach favours shifting decision-making to individual resource managers as it is thought that this will improve decision-making and provide opportunities for improved outcomes.

In this context, it is valuable to use attitudinal and public opinion data in river basin management. It provides insight into the likelihood of resource managers at the lowest level of decision-making making improved decisions.

In this section, we examine the attitudes, beliefs, values and social norms relevant to the individual resource manager in a river basin management setting. In any decision, there are usually a range of options available to a decision-maker when selecting a particular course of action. A psycho-sociological approach suggests that these choices are determined by not only the presence of institutional and biophysical constraints, economic constraints and opportunities in a river basin decision setting, but also by the decision-maker's perception of these opportunities and constraints and about the perceived outcomes that flow from the decision. Hollick (1990) interpreted this as the 'decision space', the realm in which a decision-maker operated and which was determined by his/her perceptions of possible outcomes.

When dealing with subsidiarity, it is important to recognise and understand individual decision spaces, and use them to guide local adoption of best management practices in river basin management. In this way, bottom–up can meet top–down approaches.

3.4.1 Attitudes, beliefs, values and social norms

Four psycho-sociological variables which affect decision-making in river basin management at the individual and corporate level are *attitudes*, *beliefs*, *values* and *social norms*. These have occupied a central place in the understanding of decision-making in natural resources management (Mitchell 1989). They have been used to interpret decision-making as a psychological process; that is, decision-making is concerned with the methods people and organisations employ to make choices about natural resource use.

Psycho-sociological variables guide the direction of decision-making by forming presuppositions from which a person operates. The presuppositions, or fundamental beliefs, shape a decision maker's perception of the world. In a natural resource management context, these beliefs predicate action, forming the basic values that induce valences on the means and ends within a decision situation. Feather (1982) maintained that these valences could better predict behaviour than either attitude or expectation alone. That is, if valences are measured through attitude scores, and when information about a person's expectations of the consequences of an action is included, then behaviour could be predicted. This premise is the key to understanding resource managers behaviour in a river basin. It suggests that decision-making comprises two components, a belief component and an outcome evaluation component, and that both need to be addressed if decision-making is to be used as a predictive tool to understand decision behaviour in natural resource management.

What are attitudes, beliefs, values and social norms? '*Attitude*' has been defined in many ways and has been the most widely used of psychological concepts. A well-accepted definition is that of Fishbein and Ajzen (1975) where attitude is 'a learned disposition to respond in a consistently favourable or unfavourable manner with respect to a given object'. Older definitions of the concept linked attitude causally to behaviour, but more recent definitions suggest that this linkage is more obtuse. There is considerable disagreement on the structure of attitudes, their formation, what changes them and how they influence on subsequent behaviour (Bardecki 1984). However, one framework to interpret the meaning of attitude is shown in the adjacent box.

> "An attitude can be likened to a miniature theory of science, having similar functions and similar virtues and vices. An attitude, like a theory, is a frame of reference, saves time because it provides us a basis for induction and deduction, organizes knowledge, has implications for the real world, and changes in the face of new evidence. A theory, like an attitude, is a pre-judgment; it may be selective and biased, it may support the status quo, it may arouse affect when challenged, and it may resist change in the face of new evidence. An attitude, in short, may act in various degrees like a good theory or a bad theory, and depending on what kind of theory an attitude acts like, may serve one function better than another" (Rokeach, quoted in Bardecki (1984, p. 69)).

There are four components of attitude: attitudes are affects; they are relatively consistent (transparent in interpretation); they are predispositions they are organisations of *beliefs* concerning an object. This last characteristic was developed by Rokeach (1979) to distinguish '*attitude*' from '*belief*'. He suggested that two people may respond in a similar way to an object but have quite different interpretations about the characteristics associated with that object. That is, their similar evaluations may be based on two different set of beliefs about the object. Consequently, attitude is defined as an evaluation of an object (similar to the Fishbein–Ajzen definition above), while '*beliefs*' represent the information the individual has about an object.

> "Values . . . have to do with the modes of conduct and end-states of existence. To say that a person 'has a value' is to say that he/she has an enduring belief that a specific mode of conduct or end-state of existence is personally and socially preferable to alternative modes of conduct or end-states of existence. Once a value is internalized it becomes, consciously or unconsciously, a standard or criterion for guided action . . ." (Rokeach 1979).

Rokeach (1979) developed the relationship between beliefs and attitudes, indicating that a person's beliefs, attitudes and values should be considered as a total system, the '*weltanschauung*', the totality of one's outlook on life, society and its institutions. This total system served the function of maintaining and enhancing self-conceptions that are concerned with issues of competence and morality, and that are derived in a large part from societal demands. Rokeach maintained that a person's beliefs were assumed to involve expectancies about the outcomes of holding those beliefs. An attitude was therefore a relatively stable organisation of beliefs around an object or belief that predisposed a person to respond preferentially. The concept of an attitude also incorporated discriminatory responses.

'*Values*' can be distinguished from attitudes and beliefs. They can be considered as a sub-set of beliefs, being those most resistant to change and likely to be used as an evaluative basis for the formation and maintenance of many other beliefs. People are believed to hold only a few values, what Rokeach called 'terminal values' and these form the basis of a person's world-view. These operate as foundational values in that they act as the basic premises from which people build their attitudes towards a particular item, event or issue. When an individual faces a situation where two or more values come into conflict, these foundation values are used attempt to resolve conflict and a value system comes into play. The individual's value system could be considered as a set of rules for decision-making. This is relevant to local applications of river basin management. It suggests that in any decision setting, the number of individual values may be small and river basin management programmes should target these values.

'*Social norms*' are values, attitudes and beliefs that are held in common amongst a group of people. Social norms can be defined as the degree of consensus that exists amongst a group of people, even if this consensus is non-voluntary or unintentional.

Social norms about IRBM are those sets of attitudes, values and beliefs held by a group of people sharing common experiences from their similar river basin location. However, this definition may prove inadequate, as social norms are probably determined by a number of other factors, including the influences of social interaction networks, family beliefs, the media and political beliefs, rather than relying on one simplistic causal factor, river basin location.

The use of social norms as a tool to measure the strength of similarity in community attitudes towards river basin management could be flawed by the lack of precision in determining who is the 'community' for whom social norms have been determined. The community as defined by its river basin location is misleading. People living in a river basin may not even be aware of their location in a specific basin, so this may require large-scale programmes to raise awareness and lead to greater ownership of and engagement in river basin management action. Stone (1989) observed that community interaction networks in floodplain management extended beyond physical limits of floodplains and referred more to the degree of social interaction between individuals and their families, work associates and a plethora of other social contacts. Similarly, in this book, social norms of the community are assumed to be of a wider spectrum of people beyond those generated by river basin residents.

The strength of similarity of social norms is also important, for it determines the degree of commonness of these psychological dimensions held by individuals within a social group. It also reflects group identity and a sense of belonging individuals have to local communities. These are critical concerns. If differences of opinion emerge between an individual and the social group with whom he/she acts, then one can define that person's attitudes, values and beliefs as being divergent to the group. This divergence can be quantified in terms of a measure of belief strength from the prevailing social norms, by motivation towards compliance with or individual evaluation of social norms. If a number of people hold divergent views within a community, it can be assumed that community identity is weakened, together with a sense of identity with the group and a sense of ownership of local resource management problems.

Behaviour is defined as those activities that result from decision-making. This is a function of attitudes, including those pertaining to social pressures and constraints, weighted according to their relative importance. Rokeach (1979) proposed that there are a limited set of values and attitudes a person holds in any behaviour situation, so any study of behaviour would require the selection of a discrete, appropriate, relevant but probably small number of values and attitudes to investigate behaviour.

3.4.2 The Fishbein–Ajzen theory of reasoned action

The Fishbein–Ajzen theory of reasoned action offers a framework that has conceptual clarity (Ajzen and Fishbein 1980), and is useful to river basin management. It provides the river basin manager with clear understanding of the attitudes, beliefs, social norms and intended behaviours of river basin resource managers. Research over the past few decades provides strong support for this approach (Ryan and Bonfeld 1975; Shepard 1988). Applications of the model have demonstrated its widespread use as a tool to examine resource management behaviour, in attitudes about energy conservation (Brown and Macey 1982; Macey and Brown 1983), nature conservation attitudes of farmers and conservationists (Carr and Tait 1991), attitudes to constraints in US high plains irrigation practices (Taylor and Lacewell 1988), integrated pest management (Musser et al. 1986), ex-urban expansion (Smit and Flaherty 1981), long-range migration intentions (McHugh 1984), consumer beef choice (Sapp and Harrod 1989), evacuation behaviour in response to nuclear power accidents (Ziegler and Johnson 1984; Johnson 1985) and community attitudes towards pollution (Cutter 1981; Dunlap and Van Liere 1984). A review of the Fishbein–Ajzen theory of reasoned action in over 85 studies has reported that the model was modestly able to predict behaviour (Shepard 1988).

The Fishbein–Ajzen model is stated as:

$$B \sim I = (A_B)R1, (\mathrm{SN})R2,$$

Where:

B is the behaviour

I is the intention to perform behaviour B

A_B is the attitude towards performing behaviour B

SN is the social norm affecting use of behaviour B

$R1$ and $R2$ are empirically derived weights—regression coefficients of intended behaviour against actual behaviour, that demonstrate the relative significance of one attitude to another (Ajzen and Fishbein 1980).

The two components of the model, attitudes and social norms, are measured:

(a) *Attitudes.* Attitude (A) is a function of the perceived consequences of performing the behaviour and of the person's evaluation of those consequences. Thus,

$$A_B = \sum_{i=1}^{n} b_i e_i,$$

where b_i is the belief of performing behaviour B leads to consequence i, e_i is the evaluation of the outcome i, and n is the number of beliefs about performing behaviour B.

(b) *Social norms.* Social norms (SN) are the perceived response from others about a person's performance of a specific behaviour. The general subjective norm is determined by the perceived expectations of specific referent individuals or groups, and by the person's motivation to comply with those expectations. Thus,

$$\mathrm{SN} = \sum_{i=1}^{n} \mathrm{Nb}_i m_i,$$

where Nb_i is normative belief, m_i is motivation to comply with referent i (i.e. how much the person will comply with this belief), and n is the number of referents.

The Fishbein–Ajzen model is then expressed as a regression equation:

$$BI_i = R1\left(\sum_{i=1}^{n} b_i e_i\right) + R2\left(\sum_{i=1}^{n} Nb_i m_i\right),$$

where BI is the dependent variable (the intention to perform behaviour i), and $R1$ and $R2$ are empirically derived regression coefficients that weight attitudes and social norms (the independent variables) against observed behaviour. These weights are given values proportional to the influence in the prediction of behaviour. They vary according to the behaviour being predicted, the conditions under which the behaviour is performed and the personal characteristics of the decision-maker (for example, they may include age, education, previous experience and knowledge of the targeted behaviour). Fishbein and Ajzen (1975) indicated that for some behaviours, social norms (such as the expectations of family, friends, local community) were more important in determining behavioural situations than attitudinal considerations. Various applications have been developed, identifying and analysing the intervening variables and the role of attitudes in a number of resource management scenarios.

Case studies are vignettes which provide insight into a broader theme. In the next section, we look at a case study which uses the Fishbein–Ajzen theory of reasoned action to describe floodplain management behavioural intentions. The study is at the sub-basin level of management. The approach in this case study is to recognise that floodplains in a river basin form a *decision landscape* (as discussed in Chapter 2—a spatial entity over which a myriad of individual decisions are constantly being made by resource managers). The challenge is to harness these decisions for a common, basin-wide 'good', improved floodplain management. Capturing these psychosocial data and stakeholder values will improve governance: it is the place where bottom–up meets top–down.

3.5 CASE STUDY: ATTITUDES AND SOCIAL NORMS TOWARDS DECISIONS IN INTEGRATED FLOODPLAIN MANAGEMENT IN THE NAMOI VALLEY, AUSTRALIA

3.5.1 Location

The Namoi Valley is part of the large Murray–Darling Basin (1.1 million square km) in eastern Australia. The riverine floodplains of the Murray–Darling Basin (Figure 3.8) are one of the most extensive landform types of this large river basin. They have formed from the deposition of eroded materials from the Great Dividing Range of eastern Australia. The Namoi River valley is an example of these riverine plains. It is approximately 370 km long in an east-west direction, and

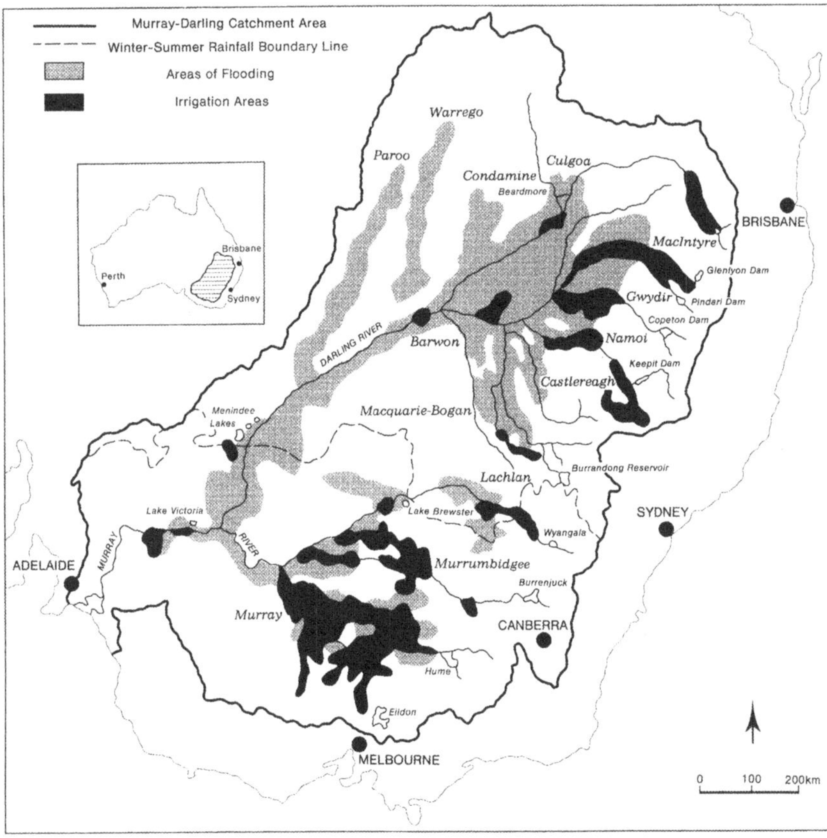

Figure 3.8 Areas of flooding and irrigation areas on riverine floodplains of the Murray–Darling Basin.
Source: Author.

170 km wide at its broadest. The length of the Namoi River is 860 km. The catchment area is approximately 41,300 square km running in a north-north-westerly direction.

Historically, the Murray–Darling riverine plains have provided significant resource use benefits as people optimised the opportunities to use the fertile soils of the floodplains. They were also urban location sites in the early European settlement of Australia because of ready access to water resources. Since the 1960s, they have been used more intensively for agriculture than ever before, despite the hazards of recurrent flooding. The Basin itself is the focus of much food and fibre production of Australia. As the nation's heartland, its floodplains are the sites of a significant, diversified irrigated (cotton, summer crops) and dryland agricultural industry. Today, the Basin produces about $10,000 million of agricultural goods per annum and includes three-quarters of Australia's irrigated farmland (Murray–Darling Basin Commission 2001).

The Basin has become the site of major environmental degradation in Australia, including irrigation-induced salinity and water logging, wind erosion, water erosion, increasing soil acidity, soil structure decline, increasing pressures on vegetation (such as woody weed infestation and eucalypt dieback), and the alienation of agricultural land. Land degradation costs were estimated at $215 million in 1987 (Murray–Darling Basin Ministerial Council 1987), the majority of which was caused by soil structure decline in cropping areas. As cropping is widespread on the Murray–Darling floodplains, this form of degradation is common to these landforms. At the same time as degradation has occurred, new agricultural developments on the northern floodplains, such as along the Gwydir, Namoi and Culgoa rivers, suggest that the process of agricultural intensification and diversification is only limited by the availability and reliability of irrigation water supplies. Many landowners on the northern floodplains of the Basin now harvest storm and flood runoff to maximise the opportunity to diversify. It appears inevitable that these developments will lead to similar land degradation problems that exist in other floodplains in the Basin.

3.5.2 Flooding regime

Flooding on the riverine plains is an episodic event. The northern part of the Basin comes under the influence of the southern extent of the summer Monsoon, generally producing higher probabilities of summer flood events. However, flooding can occur in some of the northern valleys in any month of the year. Flood duration curves are characterised by low-volume 'peaky' floods in the upper part of the valley and long duration, large volume, sluggish flooding in the lower valley—'ponding' events where streams anastomose—declining channel volume as one travels downstream. The discussion in this case study refers to all flooding environments.

Flooding in the middle (Figure 3.9) and downstream sections of the Namoi Valley (Figures 3.10. and 3.11) create long periods of inundation and disruption to rural activities for up to 4 weeks. Financial losses from flooding are considerable, but data are difficult to obtain. The floods of the 1970s were estimated to have cost cotton producers in the Narrabri district $7.98 million (1971 flood), $13.22 million (1974) and $2.18 million (1974), with an overall flooding cost for agricultural production and rural and urban infrastructure damage estimated at $100 million in 1982 values (Laurie 1982). Data on flood benefits are difficult to obtain. Significant pasture and crop production benefits followed prolonged flooding of the riverine plain after flood events in the 1970s and in 1984, but these have not been quantified. The benefits of such events are thought to be substantial by local farmers (Hooper 1994), allowing dryland and irrigated cropping and pasture production to reap substantial rewards after a flood event. Rainfall reliability is low, creating a highly stochastic environment for water supply reliability. The response by floodplain landowners and relevant agencies has been to seek a more reliable water supply situation by the construction of upstream reservoirs and on-farm storages. However, this has not ensured supply reliability. Despite the hazards of flooding and

Figure 3.9 Flooding in the Namoi Valley on the Breeza Plain, 1984.

Flooding in the mid-section of the Namoi Valley is characterised by relatively shorter periods of flood inundation than in downstream parts of the valley. Flood depths are shallow (<1 m) with flooding duration of less than 1 week. The non-perennial streams which cross the floodplains fan in ill-defined courses. Most of this area is a large depositional plain. Broadacre summer and winter cropping and livestock production dominate this agricultural landscape.

Photo source: NSW Department of Infrastructure Planning and Natural Resources.

because of the unreliable rainfall regime, irrigation agriculture developed rapidly since the 1960s, and there has been a continued demand for more land to come into irrigation development since that time.

Floodplains form within 20 km of the catchment headwater divide of the Namoi Valley, reflecting steep river gradients close to the valley boundaries. This is at a point of only 5% of total river basin length. Nearly 64% of the catchment area has slopes of less than three degrees, indicating the predominance of low elevations in the basin. There is approximately 10,720 km of floodplain land, corresponding to approximately 26% of the catchment and 40.5% of land less than three degrees slope.

Flooding constrains mainly agricultural landuse in this valley, with some serious urban flooding occurring in regional centres of Tamworth and Gunnedah, with a combined population of about 40,000. Recurring, disastrous flooding occurred in the decades of the 1970s and 1980s. While both towns are now partially flood proofed, the majority of rural floodplain land continues to be exposed to flood hazard. There have been many innovative responses to the flood hazard, with

Figure 3.10 1984 Flooding in the Wee Waa district, Namoi Valley, Australia.
This is the largest floodplain zone in the Namoi Valley. The Namoi River is the main riverine feature but several effluent streams cross the area. Due to its low gradient (approximately 1 in 400), the region resembles a large inland delta. Effluent streams, warrambools (abandoned river courses) and floodplain scours (often the site of prior stream channel traces), and the plain itself form a flooding sequence under natural conditions. The soils of the riverine plain are dominated by black and grey self-mulching clays, brown clay complexes and a variety of alluvial soils. The self-mulching nature of the black soils have produced soils of high organic matter content and fertility. Cotton production is widely practised.
Photo source: NSW Department of Infrastructure Planning and Natural Resources.

some of the first floodplain management programmes for rural Australia being developed in the lower portions of this valley in the 1970s. The evolution of rural floodplain management in the New South Wales portion of the Murray–Darling Basin owes much of its implementation to the initiatives of landowners in this valley and the efforts of the then New South Wales Department of Water Resources to develop measures to mitigate disastrous flood losses in recent decades.

3.5.3 Integrated floodplain management

The Namoi Valley is an ideal field site to investigate farmers' response to flood hazards. The high awareness of the hazard, the continuing use of the floodplains for agricultural production and the emerging awareness of the need for resource

Figure 3.11 1984 Flooding in the Wee Waa district, Namoi Valley, Australia.
The riverine plain has experienced many significant flood events, including floods in 1955, 1956, 1964, 1971, 1974, 1976, 1978 and 1984. The response by the water resources agency responsible for floodplain management was to develop a strategy of floodway restoration using levees, improved flood forecasting and warning as the main floodplain management measures, restoration programmes developed since the mid-1970s have never been fully tested by a major flood event.
Photo source: NSW Department of Infrastructure Planning and Natural Resources.

conservation of floodplain resource assets (concerns about the loss of native vegetation and wetlands, and declining soil fertility) suggest this is an ideal location to analyse the opportunities for an integrated approach to floodplain management, one at a regional, sub-basin scale. This prompted a study of integrated floodplain management in the early 1990s (Hooper 1994). In this study, the attitudes, social norms and behavioural intentions of floodplain farmers were analysed to identify the degree to which an integrated approach to floodplain management was apparent, and how such understanding could improve floodplain management at the river basin scale. Critical questions asked included:

- What are the attitudes, social norms and behavioural intentions of floodplain farmers?
- Do these vary across large river valleys?
- Can this knowledge improve river basin governance with respect to floodplain management?

To answer these questions, a new approach to floodplain management was hypothesised: integrated floodplain management (Figure 3.12), (Hooper, 1994). It is defined as an interacting set of management actions by a floodplain landowner in using soil, water and vegetation resources of the floodplain, and comprises three actions:

(1) Flood mitigation decision-making—actions designed to reduce the impact of flooding, including structural and non-structural measures.
(2) Production decision-making—defined as agronomic and livestock management actions aimed at maximising farm production on the floodplain within the constraints and incentives of the current operational environment of agriculture.
(3) Resource conservation decision-making—defined as actions involved in using and conserving soil, wetlands and vegetation resources of the floodplain.

Integrated floodplain management is a sub-set of IRBM. It is a process of simultaneous or sequential decision-making by landowners about floodplain resource use in three actions. Arrows in Figure 3.12 show the links between actions indicating the interactive nature of decision-making behaviours. The traditional approach to floodplain management, however, emphasises hazard and loss reduction but ignores the socio-economic benefits of floodplain location and resource use. Furthermore, it ignores the role of the use of natural resources in the development of floodplain management programmes that aim to reduce flood hazards and mitigate potential flood losses, even if they are multi-faceted approaches to mitigating the effects of floods. This is particularly important in rural floodplain management because the use of land and water resources for agricultural or other purposes changes the hydraulic behaviour of the floodplain. The study aimed to identify the degree to which an integrated approach exists, as reflected in attitudes and social norms towards each of the three components of integrated floodplain management, and to develop policy recommendations to improve the management of riverine floodplains.

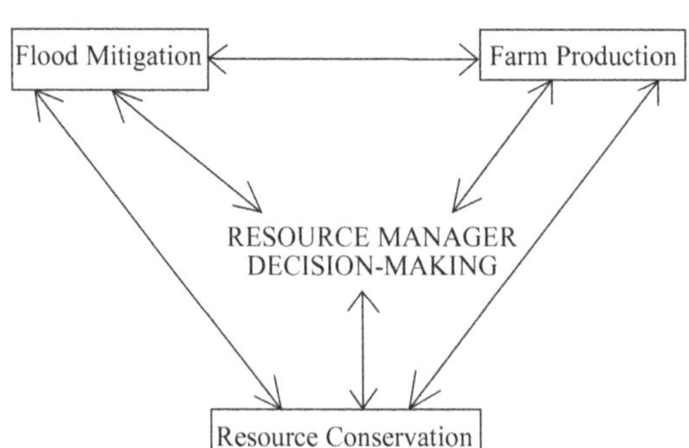

Figure 3.12 Conceptual model of integrated floodplain management.

INTEGRATED RIVER BASIN MANAGEMENT
An embracing term that includes all natural resources, resource management strategies and uses, and planning operations, interacting throughout a river basin

↓

a subset of which is:
INTEGRATED RESOURCE MANAGEMENT IN CATCHMENTS
Focusing on management of a nominated set of natural resources, resource uses and their interactions within a river basin

↓

a subset of which is:
INTEGRATED FLOODPLAIN MANAGEMENT
Above, but referring solely to management of floodprone lands, with links to other natural resources in a river basin (e.g. through catchment hydrology and flood hydraulics)

↓

A behaviour specification process follows:
NORMATIVE BEHAVIORS
A specified set of landowner actions that influence land and water resource use and management, impact directly or indirectly on flooding characteristics, and influence integrated floodplain management.

↓

A decision behaviour specification process follows:
DECISION PROCESSES ABOUT NORMATIVE BEHAVIORS
Those decision processes influencing the use of land and water resources are identified; decision behaviour is predicted through a model of intention

↓

Outcomes from the model are generated:
OUTCOME IDENTIFICATION

Figure 3.13 Derivation of an approach to examine landowner decision behaviour in integrated floodplain management.

Decision variables in integrated floodplain were hypothesised to be a sub-set of integrated catchment management and river basin management (Figure 3.13).

3.5.4 Farmer attitudes and social norms towards decision-making in integrated floodplain management

In a pre-survey, the study identified 24 normative behaviours of floodplain management, 8 for each of the 3 dimensions: flood mitigation, farm production and resource conservation (Table 3.4). These statements were interpreted as the prevailing normative beliefs about the three components of the integrated floodplain management model. The statements are a comprehensive and representative sample of an almost infinite theoretical domain of beliefs, but chosen because of their relevance to this study.

Table 3.4 Belief statements for the flood mitigation, farm production and resource conservation components

No.		Questionnaire statements
Flood mitigation		
1.	Local structural measures	Levee banks will reduce the impact of floods
2.	Upstream structural measures	Upstream dams will reduce the impact of floods
3.	Non-structural measures	Government regulation of floodplain landuse would reduce the impact of floods
4.	Coordinated networks of structural measures	A coordinated network of banks and floodways will reduce the impact of floods
5.	Upstream non-structural measures	Upstream from here, soil conservation, tree planting and better farm management together will reduce the impact of floods
6.	Non-structural measures	More irrigation development and intensive farming will further complicate flooding problems
7.	Community participation measures	Landowner groups will produce effective self-regulation of floodplain management
8.	Structural measures	Floodplain management will be more effective once the river channels are cleared out.
Production incentives/constraints		
9.	Government regulation	Farm production will continue to be restricted by government regulations
10.	Seasonality effects	Seasonal conditions will always affect farm productivity
11.	Sustainable management practices	Maintaining a farm rotation will improve productivity
12.	Financial security	More farm production will produce financial security
13.	Farming as a lifestyle	Producing agricultural goods will make for a satisfying way of life
14.	Investment motivation	The prime motivation for farming or grazing will continue to be the expected return on investment
15.	Cost impacts on viability	Increasing costs of production will impact on farm viability
16.	Debt reduction motivation	Rising farm debts will mean more farmers will have to increase production
Resource conservation		
17.	Tree clearing and environmental quality	Tree clearing will lead to a decline in the quality of the environment
18.	Wetland removal and environmental quality	Removing wetlands will lead to a decline in the quality of the environment
19.	Sustainable management practices	Stubble mulching and strip cropping will ensure the long-term productivity of this area's soil resources
20.	Sustainable management practices	Improved grazing management methods will ensure the long-term productivity of this area's pasture resources
21.	Community participation in sustainable management practices	Environmental problems will be solved better when farmers are involved in decisions on such issues

Table 3.4 Belief statements for the flood mitigation, farm production and resource conservation components (*Continued*)

22. Community participation in sustainable management practices	Voluntary soil conservation would work more effectively than legislation
23. Returns from sustainable management practices	Conservation farming will produce higher productivity
24. Community participation in sustainable management practices	Landcare groups will provide sound management of natural resources

Data on decision variables and behavioural intentions towards integrated floodplain management were collected in 11 sample sites on 3 different floodplain types (upstream, mid- and lower-valley floodplains) from a survey population of 132 landowners. Farmers' support for 24 belief statements were collected in a field questionnaire. The support was measured as support for each statement expressed as behavioural intentions.

3.5.5 Outcomes of the study

Overall, the study reported in Hooper (1994) found that:

- Decision-making was influenced by the 24 variables, and was characterised by responses to flood hazards, concerns for environmental degradation (particularly, soil degradation), structural and economic constraints on and incentives for production, and the role of agencies in floodplain management.
- Cluster analysis of decision variables revealed three dimensions to integrated floodplain management—long-term productivity, management constraints and enterprise profitability.
- Analysis of landowners' behavioural intentions revealed strong support for an agency-led approach to land regulation on floodplains, and the use of a conservation-based approach to floodplain resource use.
- There was little spatial segregation of decision-making variables, clusters of decision groups nor individual behavioural intentions across the floodplains of the Namoi Valley. This suggests that floodplain management programmes could be developed and implemented similarly across a river basin, provided that local and regional issues are addressed simultaneously.
- 'Like-minded' resource managers are most likely to be the adopters of agency programmes aimed at conservation farming, may be tentatively characterised as younger and more educated managers who are concerned about the long-term productivity of their properties.
- Evidence from landowners' attitudes indicated their intention to remain in production despite recurrent flood hazards, indicating the value of adopting a 'living with floods' approach to floodplain management.

- Decisions were based not on the immediate threat of flooding or other natural hazards, but were strongly influenced by the concerns generally held by many Australians farmers: the cost of farm production, returns on farm product, weather fluctuations and maximising the natural resource base of one's property.
- Concerns for flooding were important to floodplain managers but these concerns show little spatial dimension, nor are they related to flood experience, or other situational or psycho-sociological variables. There is no evidence to support the hypothesis that the impact of flood hazards was related to enterprise type from this study.
- Location in a river basin, particularly on its floodplains, was an important factor for 15 decision variables and these covered a broad range of concerns. However, there are no distinctive variables dominating one location in the basin.

3.5.6 What is important?

The study discovered interesting aspects of farmers' intentions to perform integrated floodplain management. Behavioural intentions were calculated for each landowner and then averaged for the total sample. A behavioural intention is the degree to which an individual is intending to perform a particular action, comply with a procedure or support a statement. Descriptive statistics are shown in Table 3.5 *t*-Tests were undertaken between observable breaks in the rank order of the data and were significant at the 1% level.

Ranked 1st: The most important intentions relate to the role of government in regulating floodplain landuse and the role of rotation practices to improve productivity. Landowners believed that the most important flood mitigation practice was the need for government regulation of floodplain landuse. They interpreted this, from additional comments gained during field interviews, as restrictions placed on structural developments and farming practices on floodplains that would affect the behaviour of floodwater. Their high ranking of this regulatory, non-structural approach to floodplain management is important to this study. This demonstrated that for landowners of the Namoi Valley, a regulatory response to floodplain development was preferred, one that regulates landowner landuse management practices. Alternatively, this could be interpreted as landowners' negative perception of a self-regulating approach to floodplain management, as it ranked 16th with a relatively low intention level (intention score = 44.6%), rather than being dependent on agency-led management procedures for floodplains.

Ranked 2nd: The next two attributes in decreasing rank order were concerned again with resource conservation issues—being involved in decision-making about natural resources management and improved methods of grazing to produce sustainable pasture production systems. Both attributes reflect a land conservation ethic prevalent across the sample population, which is noteworthy considering not all respondents were members of Landcare groups nor operated grazing enterprises.

Ranked 3rd: The next group of intentions were also significantly less important than the previous two groups and included 10 attributes, that is, nearly half the attribute set. Landowners held positive behavioural intentions towards all

Table 3.5 Behavioural intention scores for 22 significant attributes of the model of integrated floodplain management

Item no.	Attribute	Mean	St. dev.
Rank 1 Group: Preference for a regulatory approach—positive behavioural intentions			
3	Government regulation of floodplain landuse would reduce the impact of floods	73.9	19.2
11	Maintaining a farm rotation will improve productivity	73.9	19.2
Rank 2 Group: Improving natural resources management—positive behavioural intentions			
21	Environmental problems will be solved better when farmers are involved in decisions on such issues	68.6	17.9
20	Improved grazing management methods will ensure the long-term productivity of this area's pasture resources	68.4	19.8
Rank 3 Group: Threat and response factors in floodplain management decision-making—positive to neutral behavioural intentions			
23	Conservation farming will produce higher productivity	64.3	19.7
5	Upstream from here, soil conservation, tree planting and better farm management together will reduce the impact of floods	62.4	21.2
13	Producing agricultural goods will make for a satisfying way of life	62.2	21.0
19	Stubble mulching and strip cropping will ensure the long-term productivity of this area's soil resources	57.8	22.8
6	More irrigation development and intensive farming will further complicate flooding problems	56.0	21.8
14	The prime motivation for farming or grazing will continue to be the expected return on investment	55.8	22.8
2	Upstream dams will reduce the impact of floods	54.7	18.0
10	Seasonal conditions will always affect farm productivity	54.6	18.0
24	Landcare groups will provide sound management of natural resources	53.2	19.0
15	Increasing costs of production will impact on farm viability	52.6	11.2
Rank 4 Group: Diverse floodplain management variables—neutral to negative behavioural intentions			
22	Voluntary soil conservation would work more effectively than legislation	47.3	21.6
7	Landowner groups will produce effective self-regulation of floodplain management	44.6	11.4
12	More farm production will produce financial security	44.0	25.3
4	A coordinated network of banks and floodways will reduce the impact of floods	43.6	25.4
16	Rising farm debts will mean more farmers will have to increase production	38.7	18.8
8	Floodplain management will be more effective once the river channels are cleared out	36.5	18.4
9	Farm production will continue to be restricted by government regulations	31.1	14.9
1	Levee banks will reduce the impact of floods	30.8	14.9
17	Tree clearing will lead to a decline in the quality of the environment	30.7	13.2
18	Removing wetlands will lead to a decline in the quality of the environment	27.3	10.0

Scores expressed as percentages, where 100% = maximum behavioural intention.

attributes. They included concerns for improved upstream land management to reduce flooding, farming as a lifestyle, the belief in higher volumes and long-term productivity that would result from conservation farming, the perceived detrimental effect of increasing floodplain structural development on flood flows, the positive role of upstream dams as flood retention basins, threats to farming from weather variability and increasing costs, and the positive role of landowner groups. This set of attributes could be interpreted as a group of 'threat and response' factors in decision-making.

The threats are mostly external forces such as weather events (both extreme or seasonal), and in global macro-economic trends in commodity fluctuations. Yet their response is local, and self-reliant. Landowners tend to see their ability to weather these major constraints on their production systems as paramount, and the sense of satisfaction gained is rewarding, perhaps related even to the financial gains of investment return.

Ranked 4th: The remaining ten attributes had a neutral or negative behavioural intention. Surprisingly, voluntary soil conservation stands out as being a slightly negative intention. This appears to contradict the positive intentions towards soil conservation noted above, but it reflects landowners' doubts about other landowners' ability to undertake soil conservation effectively without some form of regulation.

The next set of three intentions in rank order are also less important than the previous. Landowners again displayed their lack of confidence in a community-driven approach to floodplain management and the use of networked levee and floodway systems to mitigate flooding. These results, produced some 14 years since the first structural system of floodplain management in the valley (N.S.W. Water Resources Commission 1976), reflect a general lack of confidence across the valley, not just in the riverine plains floodplains where they were constructed. Landowners do not believe that increasing the size of farm production systems produces more financial security. Their negative intention towards this attribute suggests the low regard with which it is held and that the achievement of financial security is more likely to come from reducing costs or improved land management, as indicated in their higher ranking of this item. This belief is also illustrated in the low intention towards increasing farm production from the motivation of minimising debts.

Over the entire population, channel clearing was not seen to be an effective structural flood mitigation measure. Landowners, however, were concerned with flood hazards and this was reflected in their high rankings of attributes 3, 5, 6 and 2. Another structural approach, the use of upstream dams was favoured for flood mitigation.

The lowest behavioural intentions were towards the retention of wetlands and native vegetation. This could be explained by the perceived lack of necessity to clear land on most properties in the study sample and the perceived low occurrence of wetlands in the study area. However, the findings strongly contradict the other resource conservation intentions ranked above (such as in attributes 11, 20, 23, 19 and indirectly in 5, Table 3.5). They reflect a polarity in environmental concerns, with a strong soil conservation ethic counter-indicated by a low wetland/vegetation conservation ethic. This phenomenon indicates that for these landowners there appears to be little holistic thinking towards resource conservation. They appear to make resource conservation decisions on pragmatic grounds of self-interest, reflecting their dependence on soil resources for income.

3.5.7 Does integration floodplain management exist?

The study found few distinctions between behavioural intentions between the three components of the model of integrated floodplain management. A correlation analysis between overall behavioural intentions for flood mitigation, farm production and resource conservation revealed significantly strong links between these components. It appeared that intentions towards floodplain management do not consist of three independent components, rather they are related. This suggests that integrated floodplain management does exist. Furthermore, the study also found that many interrelationships between attributes and within and between components exist, indicating that decision-making is a holistic, integrated process.

3.5.8 Decision dimensions

Further analysis in the floodplain management study was undertaken using a interviews with 47 landowners who occupied flood-prone sites in the Namoi Valley, personnel in resource management agencies, state emergency services and research workers (Hooper, 1993). Data were examined using content analysis of responses to open-ended questions regarding the type, manner and timing of decisions. The study revealed that:

- Floodplain natural resources were used simultaneously for several farm production purposes. Often, the same natural resources were used in rotation programmes and management decisions were made using a whole-farm planning framework.
- Landowners were aware of resource management issues such as wetland management, soil degradation and flooding. They impacted farm production management processes, but the degree of influence varied.
- Decision-making was not time-specific but dynamic, ongoing and responsive to changing conditions of both the biophysical and socio-economic environment, for example seasonal weather conditions and commodity price fluctuations.
- Landowners' decisions were multi-purpose. Several objectives were targeted in decision-making, but their number was usually small, reflecting landowners' practical outlook and limited time horizon for planning. Decision behaviour was a response to a range of influential variables, not just the threat of flooding. Land-use operations were geared towards production but they also operated as flood mitigation activities or other resource use activities, even if the latter were unintentional.

3.5.9 Application

These investigations suggest that agencies need to adopt an integrated approach to floodplain management, one that recognises concerns about hazard mitigation, farm production and resource conservation, and the multi-faceted nature of floodplain farmer decision-making. While a participatory approach to floodplain management with landowners is appropriate, evidence of landholder support of this concept was not convincing in this study.

Ten recommendations were developed from the study and presented to government:

(1) Floodplain management should recognise the utility gained from flooding and design floodplain management programmes that provide a degree of protection from flooding, while allowing landowners to benefit from flooding.
(2) Agencies should be more proactive by adopting a 'living with floods' approach to floodplain management. A 'living with floods' approach should use best landowner practices to manage floodplains, particularly in isolated environments. This approach will require strong landowner participation to develop best practices appropriate to local conditions.
(3) Agencies should encourage proactive planning at the local level to manage emergency situations. Improved flood warning systems in remote locations and the use of landowner flood knowledge would greatly improve current arrangements.
(4) Landowner groups, if used to formulate and implement integrated floodplain management at the local level, should be based on local resource management problems.
(5) Floodplain management programmes should adopt an integrated floodplain management approach, and when adopting an integrated approach to floodplain management, agencies and landowners should consider six critical elements of integrated floodplain management. These are the use of: agency-led floodplain regulation; best practices for sustainable resources management; a 'Learn to Live with Floods' approach; conflict resolution processes; demonstration projects and local leadership, performance indicators and empowerment.
(6) Integrated resources management programmes for floodplains should aim to address farm production, resource conservation and flood mitigation, rather than using a single issue or sectoral approach, and this approach should be applied consistently across a river basin.
(7) The promotion of 'sustainable' landuse practices on floodplains should consider the fragmentary nature of landowner decision-making regarding resource use. Property management planning should be promoted and incremental changes should be sought, rather than radical changes in farm practices.
(8) Integrated floodplain management should consider social networks between landowners when developing floodplain management practices, and recognise the strong differences in opinion about various strategies. Independent

facilitators will be required in some situations to arbitrate between landowners and between landowners and agency personnel, to resolve conflict.

(9) Floodplain management should recognise the environmental impacts of different floodplain management practices, on the biophysical, social and economic environments. Environmental impact assessment (which should include social impact assessment) should precede future landuse changes and developments on floodplains.

(10) Agencies should develop guidelines for integrated floodplain management, which are consistent with and link to other government programmes. Linkages should be developed with other agency programmes such as Landcare, Property Management Planning and the Rural Adjustment Scheme, and local government planning mechanisms. Similarly, Best practices for Integrated Floodplain Management should not be developed and promoted without reference to other farm practices.

Source: Hooper (1993).

At the time of presentation, the results were met with some interest but little real change resulted in state government policy. However, the results were of interest to the broader Murray–Darling Basin community and it is believed that they informed government policy and raised community interest in the need for improved floodplain management across the river basin. Specifically, they influenced the formation of new floodplain management guidelines in the Liverpool Plains, in the mid-section of the Namoi Valley and were used by a State government commission of inquiry to promote a regulatory approach for floodplain management under Part 8 of the then New South Wales Water Act.

3.6 NEW UNDERSTANDINGS OF IRBM DECISION-MAKING

3.6.1 Limitations of linear decision-making constructs

To recap, the approach to understanding decision-making in this chapter is to recognise three levels:

- Policy Level—high level strategy development,
- Implementation Level—the organisational level of decision-making,
- Operational Level—the individual decision-maker.

In the case study, we saw how stakeholder attitudes, social norms and values were used to inform decision processes for floodplain management. These social science elements of natural resources management performed several functions: from a minimalist perspective, they at least informed the processes of government in administering natural resources management. At a more advanced level, they acted as a healthy method of public involvement in decision processes at policy and implementation levels (refer also to Table 3.1).

The challenge is to develop decision-making processes which integrate decision-making across the three levels of governance. The approach to this dilemma is to use a reinforced bottom–up approach. This involves providing individual decision-makers with improved tools for decision-making at the lowest, appropriate level. It means giving resource managers at the operational level information which improves their ability to make informed choices. It means providing resource managers at the implementation level (such as State and Local Government agency staff) with tools to implement policy and plans which are cognizant of individual resource managers' ability to adopt best management practices. Lastly, it involves providing resource managers at the policy level (usually in government) with information about the tools and strategies which will provide more effective decision-making.

Figure 3.14 illustrates a theoretical decision framework for river basin management, focussing on the sub-regional scale. This means using a sub-basin unit of management, say a river valley, or a sub-catchment as the management unit. This would be mapped at scales of approximately 1:100,000 and refer to a meso-level scale of management. At this level, policy and river basin management planning is in place, stakeholders at all levels are engaged in decision processes and an informed populace is engaged in implementing local best management practices congruent with an overall river basin management plan.

The framework is encompassing—recognising that individual action is a response to higher level (Policy and Implementation level) actions. Individuals respond to the choices available at the time from a range of perceived alternatives.

But is this realistic? Is there such sequential decision-making? Does feedback occur and more informed decisions result? Figure 3.15 suggests that decision-making in river basin management is more likely to be episodic, opportunistic, complex and driven by the media, political agendas and pressure group interests. High-level (policy) initiatives across government or major shifts in societal behaviour may be the most influential drivers of change.

As a result, a new explanation of decision-making in natural resources management have emerged—the use of the adaptive management paradigm. There is growing recognition that this approach should be applied to river basin management, for example in the Missouri River Valley (National Science Council 2002).

3.6.2 Towards an adaptive management paradigm

This chapter included a review of decision-making in natural resources management. It demonstrated that there is a substantial body of research and practice developed since the 1960s. It emanated from the behavioural revolution in geography, particularly the focus on 'man–land' interaction in early studies. The general models of decision-making formed the basis of a new paradigm of decision-making in natural resources management. They emphasised that psycho-sociological variables, not just the prevailing concept of the 'bounded rationality of economic man' must be addressed if the process is to be understood adequately.

The dynamic, on-going nature of decision-making is also important, implying that 'rational man', a concept popular in the earliest studies of resource

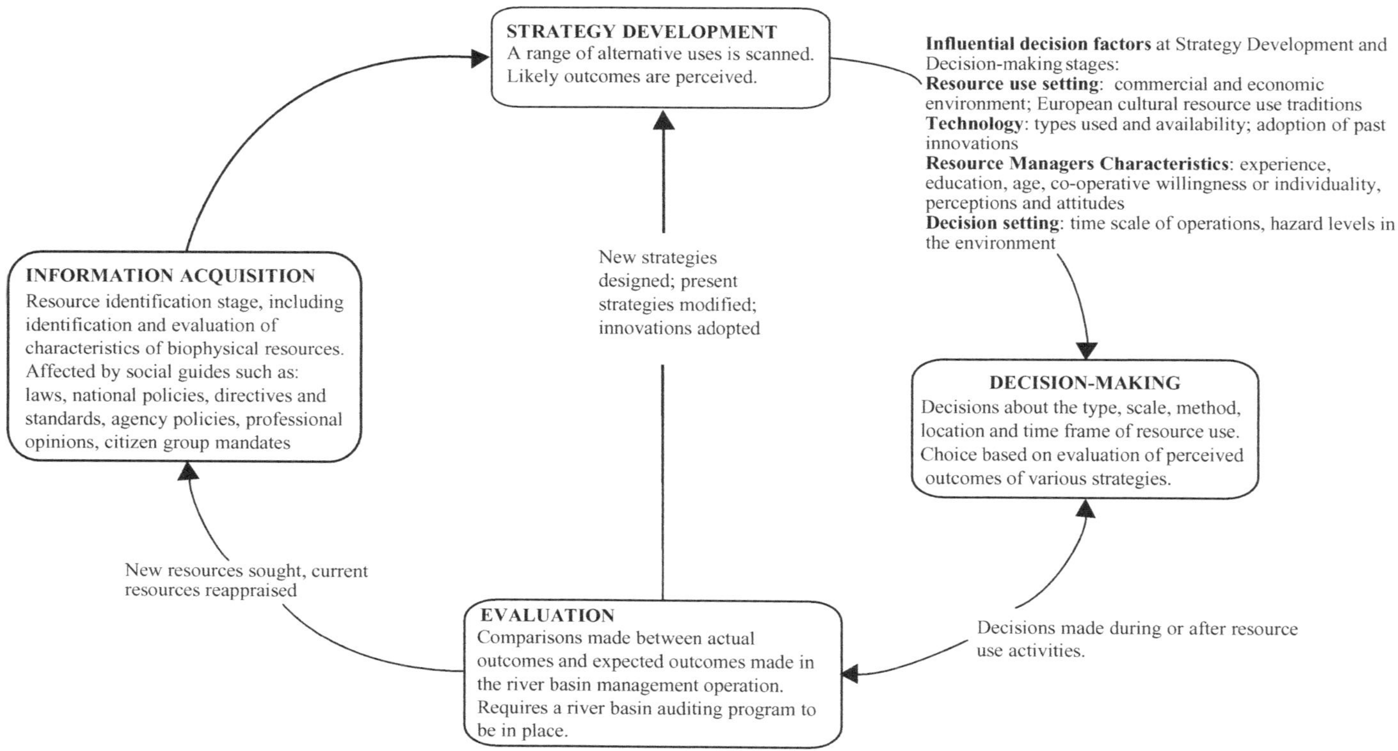

Figure 3.14 Theoretical decision-making model in river basin management.

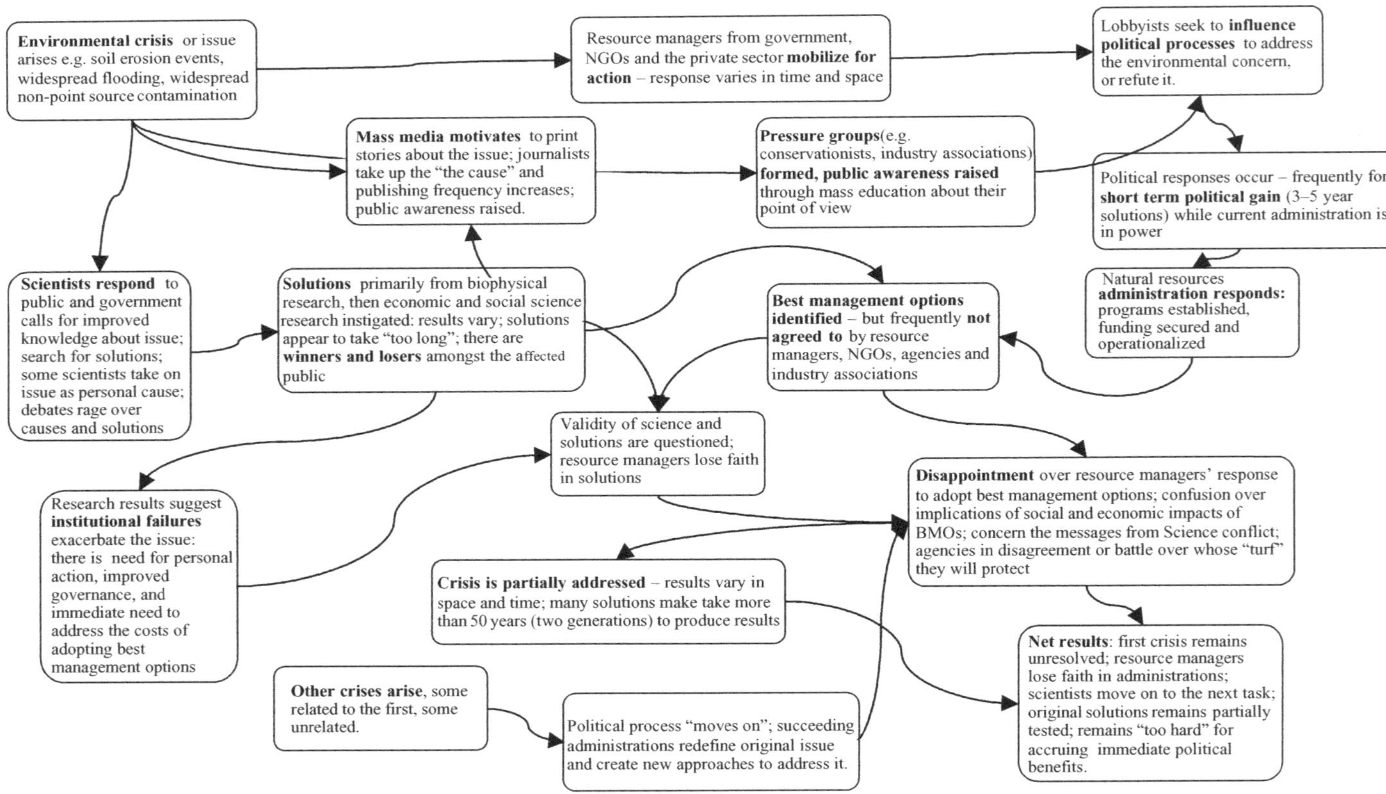

Figure 3.15 The 'messy' decision-making environment in river basin management – is this more like the real world than Figure 3.14?

management denied a complete understanding of the ways resource managers cope with natural hazards, including changing evaluations of resources and economic conditions and non-quantifiable influences on the decision-making process. This approach is more than an extension of the 'bounded rationality' approach, but a radical formulation of what decision-making involves. Over three decades ago, Bradley (1973) described this approach as an 'integrative' model of decision-making. It follows an ecological approach to the management of natural resources, suggesting that human response to choices offered by resource-use opportunities was adaptable to ever-changing conditions in both the biophysical and socio-economic environments. He stated that this type of model,

> is an attempt to evolve from the relevant insights in behavioral theory, economics, and political studies a regulatory system similar to the systematic interrelationships that exist in ecosystems of the natural environment. In the past, the belief persisted that an increased technological capacity to alter the environment brought increased control. This belief, far from dead, is a manifest delusion.
>
> (Bradley 1973, p. 300)

This statement predicts the rise of adaptive management as a way of natural resources management (Holling, 1978). The failure of traditional linear (even with feedback) planning and decision frameworks was questioned. Adaptive management today is now seen as a more robust approach to river basin governance, than traditional linear planning models. Adaptive management has two congruent, interdependent objectives: resource use and ecosystem sustainability, and requires the use of a inter-disciplinary approach, one that depends for its success on lateral thinking, learning from previous experience, inter-agency cooperation, strong leadership and extensive involvement by stakeholders at all levels of decision-making in a river basin.

The adaptive management approach uses (a) institutional, (b) organisational and (c) social resources as the key tools to implement the integrated, adaptive approach and is characterised in Table 3.6. Rather than seeing river basin decision-making as an iterative diagram, it is better to see it as a set of principles and characteristics which need to be applied to any river basin management setting.

In this context, effective planning requires two conditions to hold: (1) the planner has jurisdictional authority over the system being planned; (2) the planner can predict the consequences of his or her plans. In many river basins of the world, it is clear that, especially in multiple ownership watersheds (which is most of them), neither of these conditions holds. Adaptive management, in contrast to planning, is a process based on continuous feedback between management actions and theoretically guided interpretation of measurements of the evolving system and its interacting components that serves to minimise the error in our trials with complexity and unpredictability. This is accomplished by treating the anticipated effect of management actions on the system as hypotheses and rigorously testing these hypotheses through monitoring of system response. This informational feedback comes in the form of data, which the manager interprets as indicators that build 'stories' based upon theoretical knowledge and prior experience with the system (Kranz et al. 2004).

Table 3.6 Characteristics of integrated, adaptive river basin management

Characteristic	Description
Coordinated rather than amalgamated programmes of action	The approach emphasises bringing together initiatives from relevant government agencies, community groups and private sector organisations and identifies coordinating actions between programmes, functional overlaps and missing actions; a strategic plan of action (often expressed as a watershed management plan), prescribes team-based management plans that require stakeholders to work together for specific programmes
Top–down meeting bottom—up management	The approach recognises the value of both top–down (from government) and bottom–up (from local actions by community groups and local agencies) working together to achieve mutually agreed objectives
Strategic planning rather than all-embracing efforts	In any large watershed, there is always the challenge to deal with the tyranny of small decisions, and address them all at once. The preferred approach favours selecting those issues that are of paramount importance politically as well as critical to the health of regional ecosystems and the vitality of regional economies
A regional perspective	The approach recognises the need for coordinating mechanisms frequently at the sub-state or inter-state level across large watersheds and/or river basins. The scale of natural resource planning and management is meso-scale—at scales of 1:100,000 to 1:500,000. The approach brings together bioregional scale resource management with basin scale surface and provincial scale groundwater management
Adaptive, rather than linear approaches to resource management planning	The watershed management plan documents learning experiences in implementing coordinated actions; it is more likely to be a loose-leaf folder rather than a bound book, in that it allows review and refinement of coordinated actions, once they are implemented and reviewed for effectiveness, lessons are learned and ongoing actions are modified to improve outcomes

Table 3.6 Characteristics of integrated, adaptive river basin management (*Continued*)

Characteristic	Description
Holistic rather than single or multi-purpose management	The approach favours the management of several resource and environmental problems simultaneously; recognises that individual resource use systems (mines, industries, forestry operations, farms, fisheries) operate within a larger regional context and must be managed within the ecological thresholds (limits) of those systems, otherwise resource degradation will occur. The critical task is to identify both the thresholds beyond which irreversible change will occur in ecosystems and social systems and the acceptable limits of change to those systems
Reactive resource use planning	The approach uses modelling to predict the outcomes of current and potential practices, tests the possible outcomes against real experiences and reports these experiences back to the watershed stakeholder community. These 'experiments' in natural resource management are documented in a watershed management plan, to serve as a learning archive for future watershed managers
Creative/cost-effective rather than prescriptive financial management, using cost-sharing	Development of cost-sharing processes for different activities is seen as a way to leverage programme funds, increase community ownership of watershed management activities, and increase efficiency of integrated resource management investments; involves creatively mixing funds from other government programmes, industry groups, and the watershed community in an attempt to leverage national funds

Source: Adapted from Hooper and Lant (in press).

3.7 SUMMARY

This chapter discussed the role of decision-making in natural resources management: its evolution, various dimensions and focussed on the role of the individual decision-maker. The case study illustrated how individual values, beliefs and behavioural intentions regarding integrated floodplain management can be used to inform river basin managers.

An adaptive management approach was discussed. It calls for long-term commitment achieved through mechanisms of building institutional capacity and memory. Capacity and memory are improved by information gleaned from individual values at the local scale. It helps both river basin management policy development and planning processes, by informing decision-makers in government and elsewhere the real influences on the local decision-maker. This in turn allows river basin management decision-makers at higher levels to adapt their management programmes. As adaptive management is a relatively new approach, only time will tell if these efforts will provide improved river basin management outcomes.

3.8 REFERENCES

Aitken, S. C. 1991, Person-environment theories in contemporary perceptual and behavioural geography 1: Personality, attitudinal and spatial choice theories. Progress in Human Geography 15[2], 179–183.

Ajzen, I. & Fishbein, M. 1980, Understanding Attitudes and Predicting Social Behaviour. New Jersey, Prentice-Hall.

Bardecki, M. J. 1984, Wetland Conservation Policies in Southern Ontario: A Delphi Approach. Downsview, Ontario, York University. Geographical Monographs.

Bradley, M. D. 1973, Decision-making for environmental resources management. Journal of Environmental Management 1, 289–302.

Brown, M. A. & Macey, S. M. 1982, Understanding residential conservation behaviour through attitudes and belief. Environment and Planning A 12, 175–186.

Burton, I. 1961, Changes in the Occupance of Flood Plains in the United States. [No. 57]. Chicago, University of Chicago. Department of Geography Research Papers.

Burton, I. & Kates, R. W. 1964, The perception of natural hazards. Natural Resources Journal 3, 412–441.

Burton, I., Kates, R. W., & White, G. F. 1968, The Human Ecology of Extreme Geophysical Events. [No. 1]. Toronto, Department of Geography, University of Toronto. Natural Hazards Research Working Paper.

Burton, I., Kates, R. W., & White, G. F. 1978, Individual Choice. The Environment as Hazard. [No. 1]. New York, Oxford University Press.

Burton, I., Kates, R. W., & White, G. F. 1993, The Environment as Hazard. 2nd Edition [No. 1]. New York, The Guildford Press.

Burton, J. R. 1984, The Art of Resource Management. Armidale, Resource Engineering Department, University of New England.

Carr, S. & Tait, J. 1991, Differences in the attitudes of farmers and conservationists and their implications. Journal of Environmental Management 32, 281–294.

Chorley, R. J. ed. 1969, Introduction to Geographical Hydrology. Spatial Aspects of the Interactions between Water Occurrence and Human Activity. 1st Edtion. London, Methuen & Co. Ltd.

Cutter, S. C. 1981, Community concern for pollution. Social and environmental influences. Environment and Behavior 13[1], 105–124.

Dunlap, R. E. & Van Liere, K. D. 1984, Commitment to the dominant social paradigm and concern for environmental quality. The Social Science Quarterly 65, 1013–1028.

Feather, N. Y. 1982, Expectations and Actions: Expectancy-Value Models in Psychology. New Jersey, Lawrence Erlbaum Associates.

Fishbein, M. & Ajzen, I. 1975, Belief, Attitude, Intention and Behavior: An Introduction to Theory and Research. Reading, Addison-Wesley Publishing Company.

Found, W. C. 1971, A Theoretical Approach to Rural Land-Use Patterns. London, Edward Arnold.

Golledge, R. G., Brown, L. A., & Williamson, F. 1972, Behavioural approaches in Geography: An overview. The Australian Geographer X11[1], 59–79.
Gregg, F., Born, S. B., Lord. W. B., & Waterstone, M. 1991, Institutional Response to a Changing Water Policy Environment. Tucson, Water Resources Research Center, University of Arizona.
Harden, G. 1968, The tragedy of the commons. Science 162, 1243–1248.
Hollick, M. 1990, Land conservation policies and farmer decision-making. Australian Journal of Soil and Water Conservation 3[1], 6–13.
Holling, C. S. 1978, *Adaptive Environmental Assessment and Management*, Chicester: Wiley.
Hooper, B. P. 1993, Floodplain Management and Farmer Decision Behaviour. Research Report, Armidale, Centre for Water Policy Research, University of New England.
Hooper, B. P. 1994, Floodplain Management and Farmer Decision Behaviour. PhD Thesis, Armidale, University of New England.
Hooper, B. P. 1995, Adoption of Best Management Practices for Dryland Salinity. The Need for an Integrated Environmental Management Approach. Results of a Study in the Goran Catchment. N.S.W. University of New England, Armidale, Centre for Water Policy Research.
Hooper, B. P. & Lant, C. in press, Integrated, adaptive watershed management. Hanna, K. S. & Slocombe, S. D. (eds.). Fostering Integration: Concepts and Practice in Resource and Environmental Management. Oxford University Press. Oxford.
Hooper, B. P., McDonald, G. T., & Mitchell, B. 1999, Facilitating integrated resource and environmental management: Australian and Canadian perspectives. Journal of Environmental Planning and Management 42[5], 746–766.
Johnson, J. H. 1985, A model of evacuation-decision making in nuclear reactor emergency. Geographical Review 75[4], 405–418.
Kates, R. W. 1962, Hazard and Choice Perception in Flood Plain Management. [No. 78]. Chicago, University of Chicago, Department of Geography Research Paper.
Kates, R. W. 1963, Perceptional Regions and Regional Perceptions in Flood Plain Management. Papers and Proceedings of the Regional Science Association II, 217–227.
Kates, R. W. 1971, Natural hazards in human ecological perspective: Hypotheses and models. Economic Geography 47, 438–445.
Kranz, R., Gasteyer, S. P., Heintz, H. T., & Shafer, R. S. A. D. 2004, Conceptual foundations for the sustainable water resources roundtable. Water Resources Update 127, 11–19.
Laurie, M. A. P. P. L. 1982, N.S.W. Inland Rivers Floodplain Management Studies. Namoi Valley, Sydney, Laurie, Montgomery and Pettit Pty Ltd.
Lynne, G. D., Schonkwiler, J. S., & Rola, L. R. 1988, Attitudes and farmer conservation behavior. American Journal of Agricultural Economics 70[1], 12–19.
Macey, S. M. & Brown, M. A. 1983, Residential energy conservation. The role of past experience on repetitive household behavior. Environment and Behavior 15[2], 123–141.
McHugh, K. E. 1984, Explaining migration intentions and destination selection. Professional Geographer 36[3], 315–325.
Mitchell, B. 1971, Behavioural aspects of water management. Environment and Behaviour 3[2], 135–153.
Mitchell, B. 1989, Geography and Resources Analysis. 2nd Edition. New York, Longman.
Mitchell, B. 1991, Integrated Catchment Management in Western Australia: Progress and Opportunities. Nedlands, WA, Centre for Water Research.
Murray–Darling Basin Commission. 2001, Integrated Catchment Management in the Murray–Darling Basin. 2001–2010. Delivering a Sustainable Future. Canberra, Murray–Darling Basin Ministerial Council.
Murray–Darling Basin Ministerial Council. 1987, Murray–Darling Basin Environmental Resources Study. Canberra, Murray–Darling Basin Commission.
Musser, W. N., Wetzin, M. E., Reece, S. Y., Varca, P. E., Edwards, D. M., & Douce, G. K. 1986, Beliefs of farmers and adoption of integrated pest management. Agricultural Economics Research 38[1], 34–44.

National Science Council. 2002, The Missouri River Ecosystem. Exploring Prospects for Recovery. Washington, DC, National Academy Press.

O'Riordan, T. 1971, Perspectives on Resource Management. London, Pion.

Pigram, J. J. 1972, Resource appraisal and resistance to change. Professional Geographer v[2], 132–136.

Rokeach, L. R. 1979, Understanding Human Values: Individual and Societal. New York, Free Press.

Ryan, M. J. & Bonfeld, E. H. 1975, The Fishbein extended model and consumer behaviour. Journal of Consumer Research 2, 118–136.

Saarinen, T. F. 1966, The Perception of Drought Hazard on the Great Plains. [No. 106]. Chicago, University of Chicago. Department of Geography Research Paper.

Saarinen, T. F. 1969, Perception of Environment. [No. 5]. Washington, Commission on College Geography, Association of American Geographers. Resource Paper.

Saarinen, T. F., Sell, J. L., & Husband, E. 1982, Environmental perception: International efforts. Progress in Human Geography 6, 515–546.

Sapp, S. G. & Harrod, W. J. 1989, Social Acceptability and Intentions to eat beef: An extension of the Fishbein-Ajzen model using reference group theory. Rural Sociology 54[3], 420–438.

Shepard, B. 1988, The theory of reasoned action: A meta-analysis of past research with recommendations for modification and future research. Journal of Consumer Research 15, 325–343.

Simon, H. A. 1957, Models of Man: Social and Rational. New York, Wiley.

Slovic, P., Kunreuther, H., & White, G. F. 1974, Decision processes, rationality, and adjustment to natural hazards. White, G. F. (ed.). Natural Hazards: Local, National, Global. New York, Oxford University Press.

Smit, B. & Flaherty, M. 1981, Resident attitudes toward exurban development in a rural Ontario township. Professional Geographer 33[1], 103–112.

Stone, S. 1989, New Processes for Public Participation in Water Use Planning. Rural Water Commission of Victoria, Melbourne.

Taylor, J. G. & Lacewell, R. D. 1988, Economic experience and perceptions of U.S. High Plains farmers. Journal of Environmental Management 26, 261–275.

White, G. F. 1961, Choice of use in resource management. Natural Resources Journal 1[23], 40.

White, G. F. 1963, Contribution of geographical analysis to river basin development. Geographical Journal 129, 412–436.

White, G. F. 1964, Choice of adjustment to floods. [No. 93]. Chicago, University of Chicago, Department of Geography Research Paper.

White, G. F. 1970, Flood-loss reduction: The integrated approach. Journal of Soil and Water Conservation 25[5], 172–176.

White, G. F. 1974, Natural Hazards: Local, National, Global. New York, Oxford University Press.

White, G. F. 1997, The River as a system: A geographer's view of promising approaches. Water International 22[2], 79–81.

Wolpert, J. 1964, The decision process in spatial context. Annals of the Association of American Geographers 54, 537–558.

Young, R. W., Young, A. R. M., Price, D. M., & Wray, R. A. L. 2002, Geomorphology of the Namoi alluvial plain, north-western New South Wales. Australian Journal of Earth Sciences 49[3], 509–523.

Ziegler, D. J. & Johnson, J. H. 1984, Evacuation behavior in response to nuclear power plant accidents. Professional Geographer 36[2], 207–215.

Zimmerman, E. W. 1951, World Resources and Industries. New York, Harper and Row.

4

IRBM protocols and plans

4.1 ADAPTIVE, INTEGRATED RIVER BASIN MANAGEMENT

In Chapter 3, I discussed decision-making in natural resources management and concluded that decision-making is not a rational process with linear progression. It is more likely to be episodic, opportunistic, adaptive and reactive to changing administrative, political, environmental and economic conditions in the country where river basin management is being practised. It is frequently driven by the media, political agendas and the interests of pressure groups.

There is concern that a linear approach does not reflect the practical realities of modern natural resources management in river basins, and that an adaptive, integrated approach is more appropriate. An adaptive approach provides a way in which complex resource management problems can be addressed by emphasising management flexibility (Walters 1986). This goes well beyond traditional multi-purpose planning or sustained yield approaches. An adaptive, integrated approach to natural resources management at the basin scale is more likely to occur like this:

(1) define the problem and asses the current political climate and societal demands;
(2) gather data about the natural resources management problem;
(3) analyse the data and refine the problem;
(4) model processes to identify appropriate interventions and natural resources management options;

(5) identify coordinated management actions among stakeholders;
(6) test prototype interventions and management options in the field;
(7) audit outcomes and develop learnings from the outcomes of testing interventions and management options, and derive principles and best practices;
(8) apply prototype more fully for full implementation across the basin, for example in other subcatchments;
(9) monitor and evaluate broader applications;
(10) return to beginning, refine the problem and goals where required, and provide feedback to various stages of implementation and
(11) continue iterations of stages 3 to 10 where necessary.

In a river basin setting, it takes time for a fixed agenda to emerge about what are the critical problems and how they should be solved. The learnings in Stage 7 are derived from real-life experience. They emerge when decision makers in a natural resources management process come together to dialogue management options and test them in the field. The dialogue of adaptive management, the dialectical process when stakeholders engage, is the powerful mechanism which delivers effective basin governance.

Hooper and Lant (in press), for example, cite the experience in applying Walters' adaptive environmental assessment and management approach to the development of water quality management plans in Victoria, Australia, using a computer simulation technique. Here, it was found that it was not the computer modelling procedure (the software), rather the modelling workshops (the dialogue), which were of primary importance. Such workshops were a highly efficient medium for accumulation of information about the system and require participants to focus clearly on problems and achievable solutions from the outset (Ewing, Grayson and Argent, 2000). This experience was echoed by a similar US experience undertaken by the author in Illinois, USA. Here, Wes Seitz (University of Illinois-Extension, personal communication, 1989) maintained that more was learnt and agreed about shared governance when trialling local watershed management models in farmer café workshops, in Illinois, USA, than running the modelling algorithm. The importance of dialogue is discussed further in Chapter 7.

4.2 A PROTOCOL FOR ADAPTIVE INTEGRATED RIVER BASIN MANAGEMENT

4.2.1 The use of protocols

IRBM requires many actions by many groups, particularly natural resources management agencies in government, to bring about effective governance. The following process is one based on an adaptive, integrated approach. The aim is to include the full range of stakeholders in a river basin management.

Effective governance involves the development and ownership of a process for implementing a range of river basin management activities within the context of agreed, overall goals and objectives. The procedure includes identification, selection, testing and promotion of best management options for specific areas of a river basin, usually at the sub-catchment level. Each stakeholder needs to own this process and be able to relate to and undertake mandated tasks and responsibilities. This is difficult to achieve in practice and the challenge is to translate this approach into action. When stronger ownership is achieved and specified activities are understood within a broader management plan, implementation of integrated river basin management is more likely to succeed.

4.2.2 What is 'best practice'?

Best practice is defined as the range of on-ground actions and institutional arrangements required to address a resource or environmental degradation problem. Best practice is determined by resource managers: it is practice jointly agreed to by individuals, practitioners, government agency staff and business practitioners. It is presumptuous of any agency, organisation or individual to identify and promote these practices without a well-developed iterative procedure that engages endusers.

The following protocol for IRBM includes identification, selection, testing and promotion of best management options in specific areas of a river basin by endusers, within a whole basin context. It requires a river basin organisation and a set of institutional arrangements be used to engage endusers, resource managers, government and researchers to develop, implement and review an agreed River Basin Management Plan. The content of such a plan is specified in Section 4.3.

This protocol was originally developed in a project which combined integrated catchment management with new water treatment technologies to develop and implement a plan of action for water quality management (Hooper 1995). The issue being addressed was declining water quality in a rural catchment which supplies water to a local town. The approach was extended to urban catchments on the periphery or a large metropolitan city (Sydney) which supply water to a supply system servicing 4 million population (Hooper 1998).

In a natural resources management context, protocols are established methods used by agencies (such as departments of environment, departments of natural resources, etc.) and the private sector entities (such as natural resources consultancy firms and engineering consultancy firms) for asset management and the delivery of goods and services. They are established by industry-based evaluation—professional peer review which authenticates the validity, the reliability and the value of practices. The evaluation is built on practical field experience which refines the protocol to provide industry 'best practice.'

As IRBM becomes more widely practised, it has become apparent that there is neither an accepted 'best practice' nor a specific 'protocol' that is known to work in all situations. This is particularly important when there is disagreement and confusion about what 'shared governance' involves. There are varying environmental, socio-political and economic conditions within and between sub-catchments (in an intra-national setting) and nations (in an international setting), and these conditions preclude a prescriptive protocol. These varying conditions will demand different responses and adaptations of the following protocol.

4.2.3 A river basin management protocol

Step 1: Establish management and advisory entities

(a) *Establish a river basin management organisation (RBO).* This will include representatives of State and Local Government agencies (national setting) or national government agencies (international setting), the major private industries in the river basin, the representatives of voluntary community groups and concerned citizens from the river basin. The core task is to develop and guide the implementation of a River Basin Management Strategy and a River Basin Management Plan.

The RBO is not a local or regional government organisation—rarely do provincial governments exist, except in France and Italy. Rather it is a new entity with defined powers and functions which takes a river basin scale management approach and through political processes (influence) and some mandated powers (perhaps water allocation) coordinates the actions of local and state government departments. That is, their role is more than advisory. Examples of such organisations were discussed in Chapter 2 and include the International Joint Commission (Canada and the USA), the Mekong River Commission (Vietnam, Thailand and other Southeast Asian countries), the Chesapeake Bay Commission (USA) and the Murray–Darling Basin Commission (Australia).

(b) *Form a Stakeholder-based advisory group.* The regional RBO is advised by an independent advisory group. The management of water quality is best undertaken by a stakeholder advisory group, a government–private sector–community organisation comprising representative river basin landowners, relevant State government agencies, local government councils, local water supply authorities, and other groups claiming an interest in land and water management. The advisory panel will be supported by a technical committee providing advice on engineering, ecological, economic and social aspects of management options.

The committee's role is to provide direction regarding land and water resources management and to advise on major river basin issues and problems. The committee can also provide detailed information to assist quantifying issues, prioritising issues, identifying options to address issues, and advising on the development and implementation of a monitoring system.

Step 2: Prepare resource inventories

The next task is to prepare an inventory of resource stock. It is impossible to manage a river basin's natural resources without clear understanding of their current condition and the trends in resource quality and resource use. This will involve creating an inventory of relevant information: of both terrestrial and aquatic natural resources, of past research and other studies and creating databases which can be used to monitor the trends in quality of the river basin's resources. The databases will include information on biophysical characteristics of the river basin (e.g. soils, topography, water yield); land-use practices; adoption levels of best management practices at individual farm, forest and industry level; water quality data; nutrient source data; number and location of point sources of pollution; current status and trends in the economic viability of resource use practices; a demographic profile of the river basin; trends in population growth and composition and indicators of social well-being and social sustainability.

The assembled datasets will require spatial presentation to allow clear understanding of geographical features across the river basin. Geographic information systems (GIS) provide a useful method to collate and display the data. The use of GIS is described in Chapters 5 and 6.

One option to display land resource information is to use an ecological approach to land classification, which allows the river basin manager to assemble data into land units and land systems. The land system/land unit concept (Christian 1968) is a land evaluation procedure which can be used for this purpose. This ecological approach recognises a river basin as an integrated ecological system and uses ecological characteristics of the land surface to delineate land types, and to assess land-use alternatives both according to the potential of the land for fulfilling that use (capability) and the desirability of using it for that purpose (suitability). Land classification and land evaluation techniques are used in many applications for urban and rural land-use planning (Christian 1968; Cocks et al. 1983; Ive and Cocks 1983; Mitchell 1991).

An example of the display of this information is shown in Figure 4.1.

Step 3: Initiate appropriate studies and develop a monitoring system

There are often significant knowledge gaps in the understanding of natural resources management, so specific studies are required. The original study used for this protocol was based on the issues of water quality management in the Malpas reservoir (a rural community's water supply reservoir) and its catchment (a sub-catchment of the Macleay River Valley in eastern Australia (Hooper 1997)) (Figures 4.2 and 4.3). At this step of the protocol, several other studies were recommended including investigations into

- the nutrient sources of Malpas reservoir;
- catchment management planning for the entire catchment upstream and downstream of Malpas Reservoir;
- operational management of all water treatment and distribution facilities;
- emergency management protocols related to water quality incidents;
- the ecology of the Malpas reservoir and the upstream and downstream riparian zones;

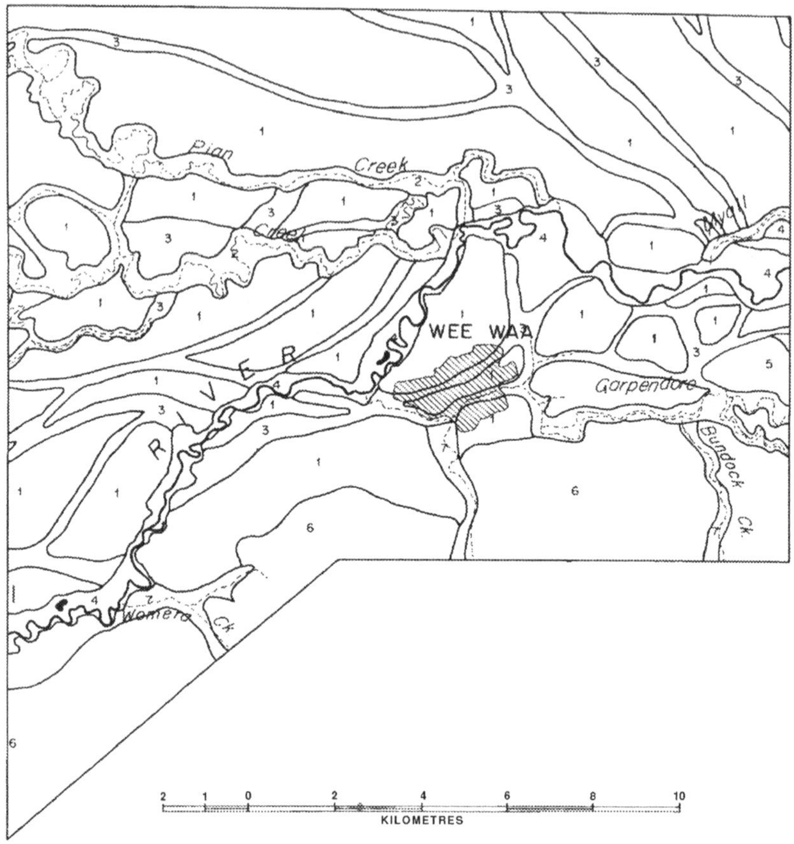

Figure 4.1 Sample land system/land unit map of the lower Naomi Valley, Murray–Darling Basin, Australia.
Source: Author

Definitions: A land system is an area or group of areas with a relatively uniform climate which are distinct from the surrounding terrain and throughout which there is a recurring pattern of topography, soils and vegetation such that the variation in these parameters is predictable. A land unit is a group of related landform elements within a particular landform relief unit that are almost uniform in their soils and vegetation, and hence in aerial photographic patterns. As land systems are integrated ecological units of land, they are useful to display resource use constraints and opportunities and specify best management options for each land management unit (Christian 1968).

Key:

NAMOI RIVERINE PLAIN
1. Clay plain
2. Effluent stream channel
3. Cross-flow area (natural floodway
4. Dissected low floodplain and Namoi River channel
5. Prior stream formation

PILLIGA SCRUB OUTWASH PLAIN
6. Board alluvial plain
7. Sand creek bed

NANDEWAR HILLS
8. Footslopes

- the role of appropriate reservoir management practices for water quality management;
- the costs and benefits of recommended river basin and reservoir management options and
- the impacts of recommended upstream land management practices on landowners, and the use of incentives to change practices.

This is the step to develop a monitoring system and includes developing appropriate multi-dimensional indicators of river basin health, water quality and operational management to provide constant information about the direction of changes resulting from use of the protocol. From the collection of these data, resource managers can prepare baseline statements of resource condition and trend and resource use behaviour to be used in monitoring programmes.

Step 4: Prioritise the issues

Each issue should be assessed in terms of magnitude (or size), and then in terms of importance (or significance) (Figures 4.2–4.4). There is also the need to predict the outcomes of different land and water resources management plans and programmes, using decision support systems that engage resource managers. These include multi-objective decision support systems linked to GISs (see Chapter 4) that engage practitioners in different sub-river basins to determine the impacts of various sub-river basin management options. The output of using decision support systems allows river basin managers to set targets for and prioritise natural resources management programmes (e.g. pollution reduction) for the whole river basin and sub-river basins.

(a) (b)

Figure 4.2 On site discussion is a valuable method of ensuring stakeholders appreciate the range, intensity and breadth of water resources management issues in a river basin management. Here, two field site meetings are shown. (a) Discussion on water quality management methods with a visiting Chinese delegation in the Malpas Catchment, part of the Macleay River Valley, Australia (where the protocol was originally developed) *Source:* Author and (b) inspection of water conservation measures in rural Andhra Pradesh, India. *Source:* Andhra Pradesh Water Conservation Mission.

Figure 4.3 The Upper Macleay Valley of eastern Australia. Water resource management issues in such a sparsely settled landscape focus more on non-point source pollution prevention and native vegetation prevention than those of highly urbanised environments below. *Source:* Author.

Figure 4.4 Issue prioritisation in metropolitan regions is a daunting task. The diverse range of stakeholders (drawn from industry, transport, commerce, civil society organisations and government) require techniques which ensure that all issues are canvassed and prioritized.
Source: Author.

Step 5: Scope the social decision system of the river basin and identify entry points

The governance of the river basin management is a critical task. This requires a clear understanding of the decision systems in the river basin available for natural resources management. There is also a need to identify where there are gaps in decision processes which need to be filled to ensure that adequate coordinated decision-making occurs in the river basin.

In this step, there is a need to characterise the social decision system: Who are the decision-makers? What are their roles and jurisdictions? What are the river basin's demographic characteristics, social networks, information flows, adoption rates of best practices and social characteristics relevant to resource and environmental management? A matrix of stakeholders and the roles and responsibilities is the first step in identifying the decision-making environment of a river basin (see Table 3.2 for an example).

The analysis of the social decision system should ideally be used to also identify the economic and social factors that cause resource and environmental deterioration. This implies a thorough analysis of the resource use history of the river basin, how resource use is changing according to the varying market, social and natural environmental conditions and the current and projected economies of the basin.

This step is not an easy task to undertake, and there has been little practical experience in undertaking this type of analysis. Social decision theory in natural resources management can be used to create this understanding. Gregg et al. (1991) demonstrated the dimensions of the social decision system in a study of water management in the USA and constructed decision rules and interventions for more effective management.

There are significant water quality and water supply reliability issues in highly urbanised regions and mega-cities of developing and newly countries with emerging economies such as India (Hyderabad, Figure 4.5). The prioritisation of issues requires not just involvement of relevant stakeholders but mechanisms in place to ensure cross-sectoral coordination with health and education programmes. In Hyderabad, the following critical water resources issues were identified in a water visioning project:

- Ensure a safe and equitable drinking water supply to all people, initiate public/government capacity building programmes to ensure the adoption of rainwater harvesting methods and to encourage the adoption of efficient water supply and utilisation techniques.
- Establish a separate working group for lake protection and management, and protect lakes from industrial pollution due to industrial effluents.
- Develop an integrated water resources plan; recycle polluted water for gardening and such purposes; promote tree plantations and encourage revival of the eco-heritage.

(Andhra Pradesh Water Conservation Mission 2003)

Figure 4.5 Hyderabad City, India – a complex social decision system exists in water resources management comprising multicultural settings, many government agencies, entrenched social deprivation and overwhelming demands for a clear, reliable water supply. *Source:* Author.

Step 6: Identify and prioritise options

This step involves the identification of river basin management options—specific management activities which are required to sustain the resource base of the river basin for current and future use. Activities will be required at several levels:

- at the individual level for specific farm or property levels;
- at the sub-basin level where there are cross-cutting issues to address, which require a broader scale of management and
- whole of river basin actions such as actions by government and institutional arrangements which will apply to all river basin managers, including cost-sharing arrangements, tax incentives, pollution abatement laws, tree preservation orders and others.

These first group of actions include a range of best management options for the individual resource manager: such as practice guidelines for individual farmers, producer organisations, local government planners, State government resource managers, extractive industries and manufacturing industries, nature conservation managers and recreation managers. These actions need to be congruent with the overall River Basin Management Plan, and designed to reduce sub-river basin pollution contributions or other negative resource management impacts. Thorough participation by all stakeholders is required to ensure that best management options are appropriate to the sub-basin by field discussions and other methods of discourse with the people who will implement them (Figure 4.6). These options need to be

Figure 4.6 The development of best management options in a rural catchment is best done by the resource managers themselves. Here, farmers in the Liverpool Plains catchment of the Murray–Darling Basin, Australia, discuss the implementation of conservation farming practices, as part of sub-catchment planning processes. *Source:* Liverpool Plains Land Management Committee.

costed, to identify the benefits and disbenefits of each option to the user community and landowners elsewhere in the catchement.

In this step, work is needed to predict the outcomes of different land and water resources management plans and programmes, using decision support systems that engage resource managers. These include multi-objective decision support systems linked to GISs that engage practitioners in different sub-catchments to determine the impacts of various sub-catchment management options. This step will involve setting targets for pollution reduction for the whole river basin and sub-catchments.

One of the potential problems encountered at this step is the economic analysis of options. This analysis may reveal beneficiaries and disbeneficiaries, most likely spread unevenly through time and space. For example, at the individual farm and sub-basin levels, the analysis of water quality management options may reveal that sub-basin location is important to those who undertake actions and consequently to those who would pay for those actions. This necessitates institutional arrangements which recognise spatial and temporal variation in outcomes and responsibilities.

Step 7: Prepare an integrated river basin management plan

This document is the core statement of action, prescribing management options across the river basin and in its sub-catchments (Figure 4.7). It is described in more

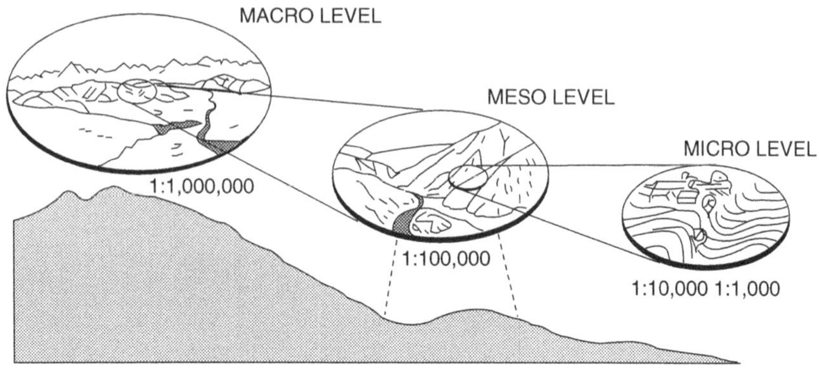

	Macro level	*Meso level*	*Micro level*
Natural system and resources	Part of a geographical zone such as a river basin or different ecological zones	Regional or local ecological resource system	Areas with relatively uniform ecological conditions
Mapping scale	>1:1,000,000	1:100,000–1:500,000	1:10,000 1:1,000
Mapping unit[1]	Provinces	Land systems	Land units, land facets
	National level	*Regional level*	*Local level and individual*
Level of decision-making	Highest political decision-making, international agreements	Province, State, District, Territory	Village, cooperative to farm, factory forest; individual
IRBM organisation	International commission	Intra-national/state basin commission/ authority/ association	Local land and water management group
IRBM document	International agreement	River basin management plan	Land and water management plan, Stormwater management plan

Figure 4.7 Scales, mapping, decision-making, organisations and documents in integrated river basin management.
Source: Adapted from Newson (1992) Figure 5.11, p. 173.

detail in Section 4.4, and includes

- an appraisal and evaluation of natural resources and their condition and trend, and content of a monitoring programme;
- an analysis of community needs;
- river basin and sub-catchment goals;
- basin and sub-catchment implementation guidelines;
- details of cost-sharing programmes for on-ground works and other actions;
- details of a monitoring programme and
- appendices which describe special river basin management issues, areas and management techniques.

[1] The units of mapping form a nested hierarchy of land units and land systems, as discussed in Step 2.

The plan will include individual and corporate actions and institutional arrangements required for sub-catchment pollution reduction. These are best management options such as practice guidelines for local councils, waste disposal industries, individual farmers, producer organisations, local government planners, State government resource managers, extractive industries and manufacturing industries, nature conservation managers, and recreation managers. These actions must be congruent with the Integrated River Basin Management Plan, and designed to reduce sub-catchment pollution contributions.

Step 8: Build a river basin information exchange programme

The purpose of such a programme is to

- inform the basin community of the management process;
- provide information to help specify best management options for sub-catchment land management units;
- describe condition and trend of the basin's natural resources;
- highlight critical management issues and where they occur;
- capture community information and knowledge about resource management.

The information exchange programme does not necessarily mean a web-base database, but can be as straightforward as a range of information exchange protocols more appropriate and affordable to developing country communities: village meetings, newspapers and radio. The programme needs to be interactive, accessible, affordable, appropriate and equitable in its distribution and use.

In Chapter 6, a prototype river basin information management system is provided, which was designed for more highly developed economies. It is easily accessible (for example, on the World Wide Web) to all key decision-making groups in the river basin. The information management system includes an interactive GIS; interactive databases which allow resource managers to contribute research and anecdotal information on best management options to a central information bank; and information on cost-sharing programmes, river basin management plans, planning zones and policies. The information management system will provide trends (through mapped environmental audits) to relay successes and failures resulting from implementing the Integrated River Basin Management Plan.

Step 9: Develop an implementation strategy and test options

This involves developing a set of resource management options and interventions. This could include actions such as

- collection and treatment of point sources of pollution;
- use of artificial wetlands to operate as nutrient sinks;
- treatment of soil erosion problems on farmland;
- treatment of streambank erosion;

- artificial destratification of reservoirs to prevent sediments from becoming anaerobic and releasing nutrients into the water column;
- improved off-take mechanisms;
- improved water treatment techniques and
- improved procedures to manage water treatment emergencies.

The testing of the options and interventions should be carefully monitored to ensure that specific goals are addressed and progress can be measured.

Step 10: Implement the strategy

Implement the Integrated River basin Management Plan—this will require:

- specification of roles and responsibilities in government, the private sector, non-governmental organisations (NGOs) and other stakeholders;
- coordination of actions of specific organisations and individuals within the river basin under a jointly agreed framework for action, signed-off by stakeholders;
- enforcement of pollution regulations under current (and needed?) legislation;
- specification of cost-sharing arrangements and
- a regular process of accounting river basin management outcomes to stakeholders.

A well-coordinated, cooperative approach is needed, driven by a committed champion of the issue, and supported by strong community representation within the advisory panel (Step 1) in the implementation process.

Step 11: Monitoring, review and feedback

Often, uncertainties result from decisions made in natural resource management programmes. It is important to adopt a feedback mechanism in the protocol. This will ensure learnings are taken on-board by the river basin organisation (RBO) and used to improve management. In this way, a continuous cycle of renewal and improvement will occur in river basin management.

The thrust of this approach is provided in Steps 6 and 10. A strong enabling environment is needed to facilitate implementation. This requires decision-makers in the river basin (government agencies, private industry, individuals and community organisations) to broker an agreed plan of action which can be 'signed off' by the RBO. This agreement will clearly specify responsibilities for action, cost-sharing arrangements for on-ground works, lines of accountability and provision of information support.

The sequence of Steps 1–11 suggests linear decision-making for river basin management. This is not the case. While Steps 1–5 form a natural progression for Step 6, it is more likely that they will take place simultaneously with the development of an Integrated River Basin Management Plan.

There is the need to recognise the Integrated River Basin Management Plan as a living document, not a shelf document, ready to collect dust. What is proposed is a document, most likely in a loose-leaf folder or published on easily changed web pages, which is continually updated by new research, more local information and lessons learned from implementing management actions.

River basin management is achieved by getting out and getting the job done within a strategic basin-wide framework. The successes and failures of using this protocol will be used to update and improve the Integrated River Basin Management Plan.

4.3 COMPONENTS OF AN IRBM PLAN

4.3.1 International and national IRBM plans

Purpose, examples and content

An Integrated River Basin Management Plan sets out goals and objectives and guidelines for management of the a river basin's natural resources. It documents data, information, procedures and mechanisms found in the protocol. The contents of the Plan will vary according to the river basin context: the type, scale and severity of natural resources problems; the level of economic development of the basin's natural resources and the capacity of current institutional arrangements and organisations to manage the natural resource management problems. There were 18 examples of river basin management plans available in November 2004 in the GWP ToolBox on Integrated Water Resources Management:

- Asia: Water saving in rice-based ecosystems (#200).
- Brazil: River Basin Committees in Sao Paulo State as instruments of participatory IWRM (#72).
- Bulgaria: Creating a watershed council along Varbitsa river (#142).
- Central and Eastern Europe: Civil Society and the Danube Basin Planning (#120).
- China: Mountain-River-Lake integrated development programme, Jianxi (#118).
- Costa Rica: How an IWRM approach would achieve better water allocation—The Lake Arenal Watershed (#10).
- El Salvador, Guatemala and Honduras: The PLAN TRIFINIO for the Upper Lempa: Opportunities and challenges for the shared management of Central American Transboundary Watersheds (#126).
- Estonia and Russia: Managing Transboundary waters in the Lake Peipsi/Chudskoe basin (#16).
- Guatemala: Towards IWRM in the Basin of Lake Atitlán (#9).
- Indonesia: A Watershed Approach to Coastal Zone Management in Balikpapan Bay (#85).
- Jordan: Amman Zarqa Basin—Using reclaimed water (#79).

- Mexico and Indonesia: Participatory Strategies for Integrated Bay and Watershed Planning and Management (#85).
- Panama: The management of the Panama canal watershed (PCW) (#5).
- Sri Lanka: Restructuring of Mahaweli Authority to an interbasin management agency (#189).
- Thailand: Budgetary procedures to provide incentives for River Basin Committees (#187).
- The Netherlands: River Basin Plan for Midden-Holland, the Netherlands (#165).
- The Netherlands and Belgium (Flanders): Cross-border cooperation for small-scale river basins (#127).
- The Netherlands: Room for the Rhine (#88).

Source: Global Water Partnership (2002). Case study numbers are included in the above listing. See website for details: www.gwpforum.org (accessed November, 2004).

IRBM plans also reflect the extent to which an enabling environment and other management instruments are in place in the basin. The following discussion outlines the generic components of a Plan.

Plan specification is best done for regions within a national and statewide policy framework. This requires moving from a single sector water planning approach to a coordinated planning approach linking land and water management. This approach contrasts with an earlier 'cookbook' approach (such as occurred in USA in 1970s) in which prescriptive rules and regulations for compliance set the agenda for water resource management agencies and utilities.

River basin management plans need to be reactive, adapting to reflect priorities and in resource management needs, and addressing the worst problems first. Plans are never 'done', they are ongoing living documents which set the stage for the next generation of plans and actions. Plans often take long time periods to produce results, decades or more, so Plans must have an inbuilt monitoring system.

The IWRM ToolBox (Global Water Partnership 2002) recommended that the river basin plans should include information on the following items.

- Physical description of the basin
- Land-use inventories
- Current water availability and demands
- Pollution source inventories
- Aquatic and terrestrial ecosystem needs
- Vulnerability to floods or extreme meteorological events
- Identification of stakeholders and mechanisms for participation
- Implications of changing land-use
- Identification of priority issues (impact issues or user requirement issues)
- Short- and long-term goals for the river basin
- Water related development scenarios, future water demands and risk assessments

- Water allocation and water quality objectives
- Strategy, measures and action plans for the achievement of goals, including sub-basin management plans
- Financing of water use and management
- Responsibility and schedules for implementation
- Mechanisms for monitoring and updating
- Annexes including specific studies such as areas of significant environmental problems

The above lists applies to both national and international river basins.

The GWP's IWRM ToolBox maintains a catalogue of river basin management agreements and experiences (listed above) and developed 'lessons learned' from implementing basin plans. The ToolBox recommended that practitioners use locally derived objectives based on a broad systems approach, and use these design principles in establishing river basin management plans (Global Water Partnership 2002).

- Use clearly defined spatial boundaries (river catchments, groundwater regions);
- Establish operational rules which reflect technical and biophysical attributes of water ecosystems;
- Employ collective-choice arrangements by engaging village and district stakeholders as well as neutral government water policy people in decision-making;
- Monitor plan and policy outcomes through water audits;
- Employ graduated sanctions, and build in conflict resolution mechanisms;
- Develop a clearly defined property rights regime;
- Separate the role of water provider from that of the regulator, to avoid conflicts of interest (how can a provider report to itself?
- Develop both demand management and supply management options, and encourage increasing water use efficiency by arrange on non-regulatory and regulatory mechanisms, particularly using current programmes to increase efficiency in irrigation and dry-land areas.

Characteristics of arrangements for international river basin management

International river basin management is fraught by a higher level of complexity—international diplomacy—than national river basin management. It is generally considered that a favourable relationship must exist between countries if IRBM is to evolve.

At the international level, an Integrated River Basin Management Plan will include international agreements which establish the need for and the institutional arrangements to manage the basin's water between countries. It is preferable to precede plans with international accords, agreements and programmes which set stable organisational conditions for the development of international river basin management. These accords allow enabling conditions to occur which can stimulate mutually beneficial relationships.

Mostert et al. (1999) maintained that there are nine dimensions of international agreements which influence the establishment of international river basin management agreements and organisations (Box 4.1). Drawing on the work of the Commission for Sustainable Development, these dimensions illustrate the capricious nature of bi-lateral and multi-lateral arrangements.

Box 4.1 Nine mechanisms for reaching international agreements that go beyond the lowest common denominator.

1. Issue linkage
Issue linkage implies that a contentious issue on which national interests conflict (e.g. upstream–downstream conflict) is linked to another issue where the distribution of (perceived) costs and benefits are the reverse. Solving such issues simultaneously can result in a net gain for all parties involved, thus overcoming the conflict of interests. The second issue might be either an RBM issue or a totally different issue, but the former is usually more effective since on both issues the same parties are involved and costs and benefits fall on the same groups.

2. Diffuse reciprocity
Diffuse reciprocity refers to countries accepting less favourable agreements in order to keep good relations and create a 'reservoir of goodwill' from which they can draw in the future.

3. Side payments
Side payments (or 'financial compensation') are payments—directly or through increased subsidies or reduced contributions—in return for a concession. Side payments will be most effective in cases of agreements affecting the economy or the finances of countries. They will be less effective when deeply held values or basic human needs are involved and can be experienced as bribery. Moreover, side payments for pollution reduction conflicts with the polluter pays principle.

4. Large geographical scope
Strict national environmental standards may limit the competitiveness of industry in a basin, but the effects are much smaller if several countries adopt similar standards for their whole territory.

5. Appealing goals/mobilising vision
Ambitious agreements can also be reached if they contain goals or a vision of the future that is attractive for large sections of society in the countries concerned (e.g. Salmon back in the Rhine, Rhine, etc.). Such goals and such a vision can act as a form of awareness raising. Moreover, they can implicitly incorporate forms of issue linkage and diffuse reciprocity.

6. Slack cutting

Slack cutting occurs when national government bodies use international agreements for introducing a more ambitious policy domestically or for promoting enforcement of existing policies.

7. Intended non-compliance

Intended non-compliance refers to the fact that countries may be willing to accept ambitious international agreements if they expect that the agreements will not be enforced. Obviously, agreements reached in this way are usually not implemented.

8. Unforeseen consequences

Ambitious agreements can also be reached if their consequences are not foreseen. Negotiators might be too confident about their national situation and assume too easily that no adaptation will be necessary. Furthermore, international courts may give unexpectedly strict interpretations to agreements. Finally, the negotiators may be inexperienced or the time to study proposals may simply be lacking, especially in the case of last-minute changes.

9. Majority voting

In some rare cases, notably within the EU, international agreements are the result of majority voting. In these cases, the more conservative countries can be overruled, at least in theory. However, the more conservative countries can link the issue to an issue where their cooperation is needed, either because unanimity is required for that issue or to obtain a majority.

Source: Mostert et al. (1999), see also Golub (1996).

4.3.2 Case Studies

Case study: the Helsinki Convention

The Helsinki Convention on trans-boundary watercourses and international lakes (Council of the European Union 1995) establishes a strong institutional framework for international agreements relevant to international river basin management. The convention established a framework for cooperation between member countries of the then (1995) United Nations Economic Commission for Europe to prevent and control pollution of trans-boundary watercourses. It recommended the use of five mechanisms to reduce trans-boundary water quality impacts: legal, administrative, economic, technical and financial measures, and used meeting of emission limits method (based on water quality criteria) to reduce pollution. Member states must work towards reducing emissions by

- ensuring that trans-boundary waters are managed in a rational, environment-friendly manner;

- ensuring that trans-boundary waters are used in a reasonable and equitable way and
- ensuring conservation and restoration of ecosystems.

The convention also recommends that member countries establish programmes to monitor the condition and trans-boundary waters within their own states.

The convention is built on three principles.

- The precautionary principle: action to avoid the release of hazardous substances must not be postponed, despite the lack of a proven causal link between the substances and the trans-boundary impact.
- The polluter pays principle: the costs of pollution prevention, control and reduction measures must be borne by the polluter.
- Water resources must be managed so that the needs of the present generation are met without compromising the ability of future generations to meet their own needs.

As well, the Convention encourages cooperation among the Riparian Parties by means of bi-lateral and multi-lateral agreements for the introduction of harmonised policies, programmes and strategies to protect trans-boundary waters. The Parties may, for example

- collect information and compile inventories on sources of pollution which have (or may have) trans-boundary impact;
- set-up joint monitoring programmes;
- adopt emission limits for waste water;
- establish warning procedures;
- carry out environmental impact assessments and
- evaluate the effectiveness of programmes dealing with this type of pollution.

The Convention requires that Riparian Parties must provide mutual assistance upon request should a critical situation arise, and must cooperate on research and development activities regarding effective techniques for preventing, controlling and reducing trans-boundary impact (Council of the European Union 1995) (http://europa.eu.int/scadplus/leg/en/lvb/l28059.htm, accessed November 2004).

This world-leading innovation in trans-boundary management frameworks is a robust set of institutional arrangements for international river basins management. The legal, mandatory nature of the convention requires real actions by signatory parties which are more likely to achieve results in trans-boundary river basin management than statements of joint intent, advisory committees, or commissions and authorities which have legal backing, regulatory capacity of funding base (see Chapter 2).

It can be used cautiously as a template for nations in other parts of the works where waters are shared. The application of this agreement is focused on water quality, a typical developed nations' perspective. So such a convention could be anticipated to be used in trans-boundary water quality arrangements in the USA such as the Great Lakes. However, it would be difficult to implement such an agreement in countries with weaker trans-boundary institutions (such as between India and Pakistan) and where political unrest or outright conflicts exist. The Helsinki Convention was predicated by other transnational agreements, relationships and administrative and political dialogues and actions. Such conditions are a precursor for such a highly developed convention to emerge.

Case study: River Basin Management Guidelines in the European Union Water Framework Directive

Historical context

River basin management began over 30 years ago through initiatives at the highest level in this region, and emerged in the expanding European Union. The overall concern was for water quality and the fragmentation of an enabling environment which precluded effective governance. The focus of the enabling environment was on legislation, and its impact on national water policies and resulting financing mechanisms within and between states.

The first attempts for improved water management began in 1975 with standards set for water abstraction from rivers and lakes and in 1980 with standards set for drinking water quality. In 1988, a ministerial seminar in Frankfurt reviewed existing legislation and identified problem areas. The result was a suite of legislative reform in 1991 including:

- The Urban Waste Water Treatment Directive, which aimed to provide secondary (biological) waste water treatment;
- The Nitrates Directive, which address agricultural nitrate pollution, and later in 1991.
- A Drinking Water Directive, which reviewed water quality standards (adopted (1998)) and
- A Directive for Integrated Pollution and Prevention Control (adopted (1996)), which addressed large industrial effluent pollution.

However, the thrust towards river basin management took hold of concerns in the mid-1990s about the need for an overall framework to manage water. Using a widespread consultation process and a 2-day water conference in May 1996, the need emerged to overcome fragmentary water policy and establish a single piece of framework legislation. This became known as the European Union Water Framework Directive, a regional sustainable development initiative. Consisting of 26 Articles and supplementary annexes, the Framework sets the following objectives of an integrated approach to water resources management.

- Expand the scope of actions to protect water to all forms of naturally occurring water in the environment, including surface and groundwater.

- Prevent further deterioration, and protect and enhance the status of aquatic ecosystems, and with regards to their water needs, terrestrial ecosystems and wetlands (Article 1 (a)).
- Promote sustainable water use based on long-term protection of available water resources (Article 1 (b)).
- Take specific pollution control measures, by reducing or eliminating discharges and emissions and losses of priority toxic substance, to enhance the protection and improvement of the aquatic environment (Article 1 (c)).
- Reduce pollution of groundwater (Article 1 (d)).
- Contribute to mitigating the effects of floods and droughts (Article 1(e)).
- Undertake measures which will result in the achievement of 'good water status' for all waters within a predetermined time scale.

Chave (2001, p. 10), http://europa.eu.int, accessed November 2004.

The implementation of the Framework is undertaken through, and is a useful example of, an integrated, cross-sectoral, coordinated approach.

- A single system of water management: river basin management, in which the emphasis was placed on international and national water resources management than focused in hydrologic units as the focus of management, not political boundaries.
- Coordination of objectives to achieve three outcomes simultaneously: protection of special aquatic habitats, protection of drinking water resources and protection of bathing water.
- Protection of surface water through ecological and chemical protection mechanisms: establishing procedures to establish chemical and hydromorphological standards unique to each body of water and protection mechanisms which ensure that minimum quality standards are reached.
- Groundwater protection with zero-tolerance pollution being enforced. This involves not a standards approach, rather one which prohibits direct discharges to groundwater, and setting withdrawal limits so that only that portion which is abstracted which does not harm ecological requirements is allowed.
- Coordination of measures: ensuring that any action meets legislative requirements in the first instance and does not work against other objectives in the water resources management setting. The Framework also establishes a system of prioritisation of objectives and resulting action programmes to establish strategically important problems first.

Source: http://europa.eu.int, accessed November 2004.

The river basin management plan

The strength of the spatial locus of integrated water resources management in the EU Water Framework, the river basin, is also a challenge to water managers. The

Framework requires that

> All the elements of this analysis (the Framework) must be set out in a plan for the river basin. The plan is a detailed account of how the objectives set for the river basin (ecological status, quantitative status, chemical status and protected area objectives) are to be reached within the timescale required. The plan will include all the results of the above analysis: the river basin's characteristics, a review of the impact of human activity on the status of waters in the basin, estimation of the effect of existing legislation and the remaining "gap" to meeting these objectives; and a set of measures designed to fill the gap. One additional component is that an economic analysis of water use within the river basin must be carried out. This is to enable there to be a rational discussion on the cost-effectiveness of the various possible measures. It is essential that all interested parties are fully involved in this discussion, and indeed in the preparation of the river basin management plan as a whole.
>
> http://europa.eu.int/comm/environment/water/water-framework/overview.html, accessed December 2004.

How is this to be done? The framework sets out the specific functions of a river basin management for both intra-nation and international (by default) river basins in the preceding list of objectives. The contents of a river basin management plan are driven by the outcomes of the analysis of resource management options. This analysis is also included in the document. Chave (2001) specifies the content of a river basin management plan as including:

1. general description of river basin (Article 5);
2. summary of significant pressures and impacts of human activity;
3. identification of protected areas;
4. maps of monitoring networks and the results of monitoring;
5. environmental objectives (Article 4) including extensions and derogations;
6. summary of economic analysis of water use (Article 5);
7. summary of programme of measures (Article 11);
8. summary of measures to implement EU legislation for water protection;
9. details of practical steps to recover costs of water use (Article 9);
10. measures taken to protect drinking water resources;
11. summary of controls on abstraction, impoundment, point source discharges and other activities (Article 11);
12. authorisations to direct discharges to groundwater (Article 11);
13. summary of measures to deal with priority substances (Article 16);
14. summary of measures taken to prevent accidental pollution;
15. details of supplementary measures needed to meet environmental objectives;
16. measures taken to reduce marine pollution (Article 11 (6));
17. register of more detailed plans or sub-basins, sectors, issues or water types;
18. summary of public information and consultations, and their results;
19. list of competent authorities and
20. contact points for obtaining more information under Article 14(1), control measures described in Article 11(3) and actual monitoring data.

The case demonstrates how the definition of IRBM (as discussed in Chapter 1) has been translated in a way which prioritises a stronger enabling environment through legislative backing—it is a directive of an international 'federation' of states (see Chapter 2). The Directive empowers and requires compliance from member states to build policy initiatives and river basin management plans. These are the core implementation tools.

The challenge of this approach is that it adds high transaction costs to some member states who are now required to work across borders and build liaisons with neighbour states who share a river basin. A further issue is that it requires data sharing for joint economic analysis of water use options. This will require significant work to establish these information sharing protocols. It is not unrealistic to expect member states to see this as another layer of administration to be established within their own governments. This attitude is not unexpected and is similar to other experiences in integrated catchment management in Australia where interstate and intrastate administrative issues abound (House of Representatives Standing Committee on Environment and Heritage 2000; Comino 2003; Healthy Rivers Commission of New South Wales 2003). However, Chave (2001) suggests that many of the EU member countries already have international agreements for shared waters management and this may not be that difficult compared to nations which do not. The Rhine and the Saar basins for example have international commissions which can be adapted to address the Water Framework Directive's requirements. The challenge will be to take older commissions (such as the Rhine, established for navigation purposes) and create new institutions able to adopt IRBM.

The EU approach to IRBM is being worked out through case study basins. These were identified in 2003 (Article 3) and at the time of writing the pilot basins endorsed by the Water Directors of the programme included: the Scheldt (Belgium, France and The Netherlands), Moselle-Sarre (Germany, France and Luxembourg), Marne (France), Shannon (Ireland), Ribble (UK), Odense (Denmark), Oulujoki (Finland), Sudalsvassdraget (Norway), Guadiana (Portugal), Júcar (Spain), Pinios (Greece), Tevere and Cecina (Italy), Somos (Hungary and Romania) and Neisse (Czech Republic, Denmark and Poland). These basins will be used to test and validate the guidance documents developed by the Water Framework to enact river basins management.

There have been mechanisms to support the integrated testing of river basins management guidelines in the pilot river basins. The Joint Research Centre of the European Commission and the Institute for Environment and Sustainability (http://ies.jrc.cec.eu.int/) have established an information exchange facility among those responsible for testing in a Pilot River Basin and those have been involved in the provision of guidance documents. The platform is implemented as document/information space, a help desk and a set of mailing lists. The last is listed as Box 4.2.

The trajectory of the basin management review process is summarised in Figure 4.8 and reflects the integration across other Water Framework Directive initiatives. The draft river basin management plans of the EU members states are

Box 4.2 Information exchange network (email list servers) for personnel engaged in the Pilot River Basins of the European Water Framework Initiative.

List name	Area covered by mail list
PRB-IMPRESS	Analysis of pressures and impacts
PRB-HMWB	Identification and designation of heavily modified water bodies
PRB-REFCON	Reference conditions for inland surface waters
PRB-COAST	Typology and classification of transitional and coastal waters
PRB-INTERCAL	Inter-calibration
PRB-WATECO	Economic analysis
PRB-MONIT	Monitoring
PRB-GROUNDWATER	Tools for assessment and classification of groundwater
PRB-PUBLIC-PARTICIPATION	Best practices in river basin planning: public participation
PRB-PLANNING-PROCESS	Best practices in river basin planning: planning process
PRB-WATERBODIES	Issues related to water bodies
PRB-GIS	Geographic Information Systems issues
PRBPARTICIPANTS	General list concerning the integrated testing in pilot river basins

Source: http://viso.ei.jrc.it/wfd_prb/enter_pie.html, accessed December 2004.

timetabled for completion in 2008, with finalisation in 2009. The current management cycle (ends 2021) and the future management cycle is scheduled for 2021–2027 (Articles 4 and 13). The value of this long-term effort will be the ability with which the programme can review the Pilot River Basins and articulate the strengths, weaknesses, opportunities and methods to overcome threats of this new approach. The development of better guidelines, politically and financially supported will allow improved basin governance.

Case study: World Bank Guidelines for Integrated River Basin Management

The development of IRBM at the international scale was further specified by Millington (1999) in advising the World Commission on Dams. He stated that an effective river basin organisation is the one which

- operates in a *stable institutional framework* that overcomes fragmentation and overlap of responsibilities, and is supported by strong and comprehensive, but flexible legislation, regulations, decrees etc. This ensures 'fairness' in basin-wide decisions and a process of accountability.

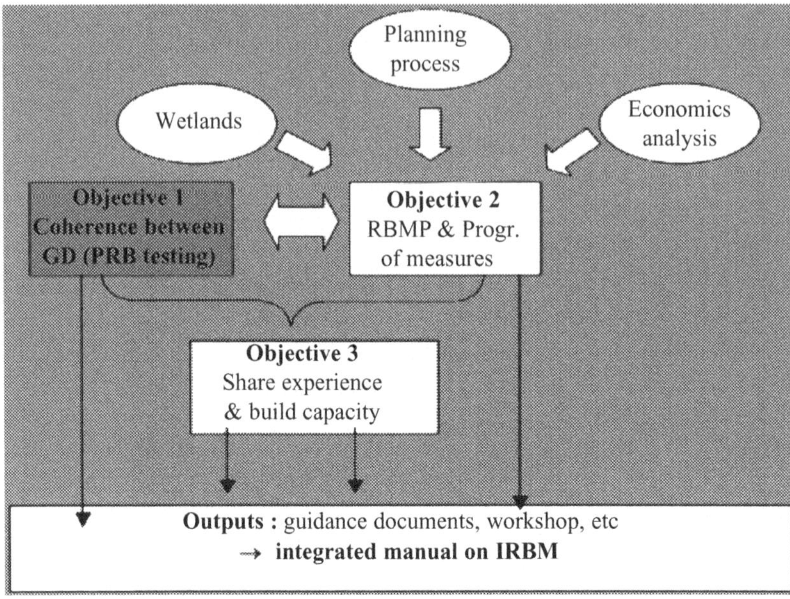

Figure 4.8 Trajectory of river basin management review process.
Source: Pilot River Basin Directory of the Water Framework Directive: http://viso.jrc.it/wfd_prb/intro.html, accessed November 2004.

- uses a *strong knowledge base* that derives from a good, uniform and comprehensive data network, systems and models for analysis, and that allows 'knowledgeable' natural resources/water management policies and strategies to be developed and implemented.
- *integrates action* across all natural resource issues, which means agencies do not find singular solutions but look at impacts and improvements across the spectrum of natural resources, and the development of regional (basin scale) natural resources management policies.
- uses *strong community awareness and participation processes* to enhance greater farmer ownership of basin scale plans of action.
- has a *strong foundation and mandate in legi*slation which clearly identifies its functions, structure and financial base and whose administration and operation is based upon a decision-making process of authority, responsibility and accountability.
- is *conceived in the reality of existing conditions*, where there are vested interests, attitudes and economic bases. Where reforms of the magnitude of river basin management are introduced or expanded, there is resistance to change and concern over infringement on administrative level and agency 'turf', so a strategic planning and implementation process based on communications, coordination and cooperation within a river basin organisation is developed.

His approach to basin management was developed further in briefing notes for World Bank staff to assist their roles in developing policy, river basin management plans and the establishment of river basin management organisations (Millington 2004). The notes were developed for both within-state (national and provincial) and international river basin settings, and drew on his experiences in the Mekong River Commission (South-east Asia), the Murray–Darling Basin Commission (Australia) and the Tarim Basin Water Resources Commission (China) (see Annex 1 except for the Murray–Darling). The draft edition was divided into five parts and 15 separate notes (Box 4.3). This framework establishes a policy, basin planning and procedural measures for river basin management.

Box 4.3 Components of the Draft Briefing Notes on Integrated River Basin Management for World Bank staff.

A. Concepts and Institutional Issues
Note 1. Background, which scopes the need for integrated river basin management; types of river basin organisations; the need to separate roles and functions (of resource managers, from those of pollution monitors and regulators and from service providers) for the clarification of responsibilities in basin management.
Note 2. Creating and Empowering a Basin Organisation, which focuses on the role of 'mutual benefit' and doing the 'right thing' by customary law to establish international basin organisations.
Note 3. Organisational Strategic Planning for a River Basin Organisation, which sets the direction, defining the priorities, planning the actions, monitoring the results.

B: Data and Information
Note 4. Water Related Data and Information Management, in which transparent, open information exchange is advocated.
Note 5. Water Related Resource Inventory, which includes good data and information on the condition of the natural resources bas, a well developed set of simulation models for testing policies, development options and projects, and a set of decision support tools to present the modelling information in a way which helps decision makers.
Note 6. Systems Modelling, which simulates the behaviour of the basin's resources in response to new policies and development options and the use of a package of decision support tools by working groups and the use of continuous staff training to maintain decision support capability.
Note 7. Notification and Evaluation of Projects, which outlines the requirement to establish notification of new projects to all basin stakeholders and evaluation techniques, including environmental impact assessment.

C: Policies and Strategies
Note 8. Sharing and Managing a Basin's Water Resources, which outlines methods of reasonable and equitable water allocation drawing on case study

experiences, the use of water accounting mechanisms, quotas, transfers and audits.
Note 9. Licensing/permitting of Water Diversions and Use, which involves setting the rules for water licensing, issuing the licenses, monitoring who uses how much water and how efficiently.
Note 10. Modern Approaches to River Basin Planning and Management, which focuses on engaging basin communities, the role of bottom–up planning and participation in local land and water management plans.
Note 11. Pricing and Charging for Water Resource Management, which outlines the role of efficient water pricing structures for both supply and distribution, and for managing and monitoring the resource base itself, and the role of independent pricing tribunals.

D: Partnerships and Awareness
Note 12. Stakeholder Partnerships, Participation and Funding, which outlines partnership building methods with peak bodies and at lower levels.
Note 13. Raising the Awareness of the Basin Community, which outlines the contents of a basin package of communication initiatives spanning education on IRBM for schools, villages, towns and the community in general.

E: Monitoring and Capacity Building
Note 14. Setting and Managing Basin Sustainability Performance Indicators, which outlines the need for sustainability benchmarks and performance indicators of river basin management, and the contents of a river basin 'status report'.
Note 15. Setting directions, informing and motivating staff, creating a vibrant, respectful organisation, which outlines the project management cycle for river basin management, organisational performance enhancement, marketing river basin management to stakeholders, and the role of leadership.

Source: Millington (2004), personal communications with World Bank staff, February 2004.

4.3.3 IRBM sub-basin plans

A sub-basin or land and water management plan (LWMP) is a tool to enact integrated river basin management and local levels. It includes priorities, actions and reporting mechanisms for implementing the overarching IRBM Plan at the lower level of a river valley or smaller catchment within river basins. Here, planning and management is at scales of approximately 1:100,000. It include sections or chapters on:

- *Context* (a review of the historical, economic, environmental and statutory context within which the Plan will work) and *Scale* (the size and level of application of the Plan statewide and in districts).

- *Engagement Processes*: Methods used to engage and use stakeholders to build the Plan; and external auditing of engagement process.
- *Water Services*: Statements of present and future needs and issues; statements of present and calculations of future requirements for water services in terms of water demand and supply, for urban, rural towns, industrial, power generation and irrigation users; determinations of supply reliability in stochastic climatic and commodity environments.
- *Other water services*: Floodplain management, salinity management, groundwater management, river management, water quality management and other key issues where appropriate.
- *Determinations of Management Options and Specific Courses of Action*: Development options, demand management options (structural, legal and economic (e.g. user pays solutions); implementation tasks, who is responsible, funding sources; coordination options with other government departments.
- *Institutional Arrangements: Structure, Governance and Functioning of basin organisations*: Purpose and scope, organisational arrangements, composition and representation, decision-making rules, funding and staffing, and authority of each government department.
- A *Monitoring Programme* to measure successes and failures of plan and provide accountability of government investment; congruence and linkages with statewide state of the environment reports and/or environmental auditing and monitoring.

4.3.4 Case Study of IRBM sub-plans: A Land and Water Management Plan in the Warangal district, Andhra Pradesh, India

Introduction

The Warangal District (a local government region) in the State of Andhra Pradesh, India (Figure 4.9), is an example of an application of IRBM at the local level. The district straddles two river basins, Krishna and Godavari. It provides valuable insight into how land and water management planning at the local level can be harmonised with river basin management and State-wide land and water resources management.

The following discussion is derived from the experience of water visioning and planning for river basin management and State water policy. It is documented more fully in (Andhra Pradesh Water Conservation Mission 2003) and (Andhra Pradesh Water Vision Task Force 2004). Much of this section comes from the latter document, a working paper used in a workshop to facilitate LWMP.[2]

The State of Andhra Pradesh has a long history of water management, but only very recent experience in integrated water resources management and IRBM. In

[2] The author acknowledges the contributions of Dr K.V.G.K. Rao, Water Conservation Mission, Hyderabad for his valuable contributions and insights to this case study.

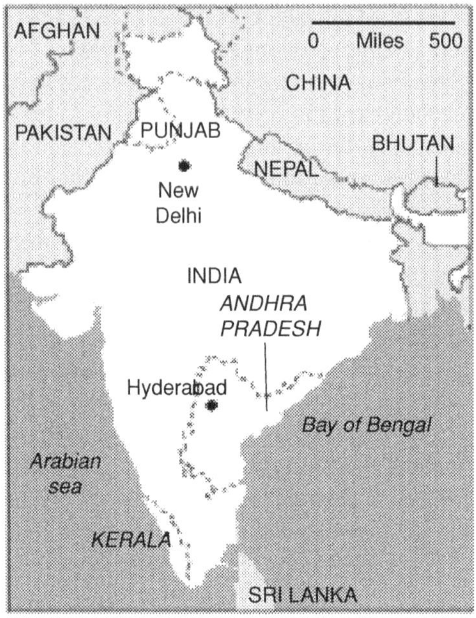

Figure 4.9 Location of Andhra Pradesh in India.
Source: Author.

fact, proposals have only been developed in 2003 and 2004 to use an integrated approach. This case study is used to demonstrate how an emerging process is being used to harness IWRM principles at the local scale in river basin management.

The discussion concludes by elaborating on how this approach to local land and water management planning in a developing country can be utilised in similar situations. In this way, it addresses some of the criticisms of transfering 'western' river basin management to developing countries (Shah, Molden and Sakthivadivel, 2003).

The Andhra Pradesh Water Vision

The development of land and water management planning in the District (a local government unit) of Warangal was preceded by a 2-year process to establish a Water Vision in the State. The Government of Andhra Pradesh established a Water Conservation Mission in 2000 with support from the Royal Netherlands Embassy, New Delhi, and produced the AP Water Vision (Andhra Pradesh Water Conservation Mission 2003).

The Andhra Pradesh Water Vision was developed in 2001–2003 and released in August 2003. It charts a strengthened course for water management. The Water Vision focuses on delivering *water security* and advocates implementation at State, District and local levels. A snapshot of the critical theme of water security is elaborated in eight key vision statements (Figure 4.10):

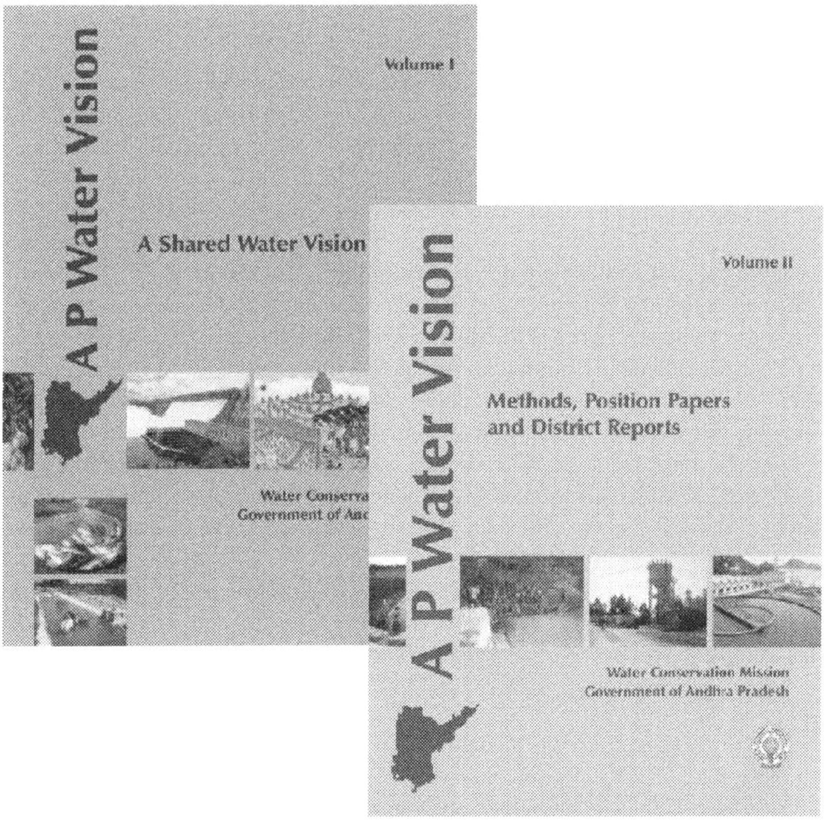

Figure 4.10 The Andhra Pradesh Water Vision: the result of a process of engagement with water stakeholders in villages, non-government organizations, water mentors. *Source:* Andhra Pradesh Water Conservation Mission.

- Clean, hygienic, accessible, affordable and secure drinking water supplies for the entire population.
- Sustainable levels of water extraction from rivers, tanks and groundwater—without jeopardising their future use or vital ecosystem functions.
- Conservation of rainwater and its efficient use for agriculture, plantations, livestock and groundwater recharge.
- An efficient, well-managed and sustainable irrigated agriculture sector—enhancing value and ensuring farming livelihoods, but also avoiding wasteful use of water. Of great importance is the efficient use of water in agriculture—maximising the return on water and the social benefits of efficient water-use.
- Mitigation of the effects of droughts, with short-term emergency responses and long-term planning.
- Prevention of the pollution of water resources used by people and livestock, agriculture and industry.
- Integrated governance of water—reflected by effective legislation, efficient government services that work in a coordinated manner, sound water information

and data sets, adequate monitoring and applied research—so that we know where we are and what options are available to us.
- Participatory water management through effective institutional arrangements. Greater concern for water management at every level—individual, community and government. Special emphasis on the participation of women and landless persons in decision-making.

Source: Andhra Pradesh Water Conservation Mission (2003).

Figures 4.11–4.15 illustrate dimensions of the Water Vision.

Recent political developments in India and Andhra Pradesh promote the delivery of government services at the district level. With the possible introduction of an AP District Planning Committees Act, there is the opportunity to implement the Water Vision at the district level, within a 5-year District Plan. Even without this Act being implemented, the current arrangements for district governance could be mobilised to implement the Water Vision, as a District Land and Water Management Plan (LWMP). This is not a static document, rather the plan guides ongoing, adaptive courses of action. The plan was also developed for Warangal district as a template for other district LWMPs.

Such plans are created in the context of and will be congruent with proposed new State Water Policy and a Water Resources Management Plan for Andhra Pradesh. These are still evolving and their creation was recommended in the Water Vision,

Figure 4.11 Collection of water vision messages for the AP Water Vision. As well as district workshops, many informal meetings were held with farmers who were most willing to provide their input: here outlining successful watershed (rainwater harvesting) methods in the Nalgonda district. Here, Dr KVGK Rao, of the Water Vision team far right, discusses watershed development with a farmer and staff of an NGO (AFPRO) in the Nalgonda district.
Source: Author.

Figure 4.12 Urban areas in regional Andhra Pradesh present unique problems of water management: burgeoning populations, poor infrastructure maintenance and only periodic supply. The answer lies in building the capacity of local water supply organisations to provide non-permeable groundwater tanks, ongoing tank rehabilitation and assisting traditional tank managers to enact their oversight role.
Source: Author.

Figure 4.13 Nagarjuna Sagar Dam on the Krishna River, Andhra Pradesh. The dam was built in the 1960s to provide major irrigation water to downstream coastal irrigation regions. The Water Vision recognises the value of the resource but also called for the need for improved efficiency of downstream irrigation techniques.
Source: Water Conservation Mission, Andhra Pradesh.

Figure 4.14 Irrigation supply channels double up for many purposes in a State with increasing water demands. The Water Vision called for securing water resources above all else, so that an adequate amount of potable water could be supplied to all and that encroachment of other supplies can be minimised.
Source: Author.

Figure 4.15 The implementation of the Andhra Pradesh Water Vision involved intense dialogue between agency staff to develop specific work plans within their own departments, determining who is responsible for actions and how coordination will be achieved: here through a workshop process.
Source: Author.

based on the principles of integrated water resources management, and congruent with the 2000 National Water Policy in India (Government of India 2000).

Methods to develop the Water Vision

During the development of AP Water Vision, several methods were used to capture 'water messages' and develop a shared vision for water resources management in the State. Over 600 water messages from many government, NGO, district and water user groups through the State and condensed into a Strategic Water Framework and Vision statements. Workshops played a key role at all levels (district and city, government and rural community) to distil directions for future water resources management.

The Water Vision in the Warangal District

The workshops in Warangal District were held in April 2002–June 2002 and November 2002–December 2002. A total of 45 persons participated in the Warangal workshops—18% from community-based organisations (CBOs), 33% from NGOs, 40% government representatives and 9% individuals. The following vision statements emerged from the workshops:

- Improve the quality/quantity and reliability of drinking water supply, and reduce the fluoride content of water wherever necessary.
- Build-up a sustainable groundwater resource, by reducing exploitation, imposing limits on industrial pollution, and constructing more rainwater storage structures.
- Promote agricultural activities that use water-efficient crops and efficient water-use techniques through strengthened Water Conservation and Utilisation Committees.
- Improve the coordination between committees and establish conflict resolution processes to reduce local disagreements.
- Initiate pilot water planning projects.
- Provide compensatory charges to aqua-farmers to encourage improvements in water quality.
- Encourage rainfed horticulture and forestry.

Characteristics of the Warangal District

Location and Water Administration

Warangal is one of 23 districts of Andhra Pradesh and lies in the north of the State (Figure 4.16) and covers 12,800 km^2. The district lies between latitude 17° 19′ and 18° 36′ North and longitudes of 78° 49′ and 80° 43′ East.

For the purpose of administration, the district is divided into five revenue divisions and 51 mandals (Figure 4.17). A mandal is a subunit of local government.

Figure 4.16 Districts of Andhra Pradesh.
Source: Andhra Pradesh Water Conservation Mission.

Warangal district is well known for its history in tank irrigation. The district is famous for its rich history. *Orugallu*—the historical name of Warangal City—was the capital of the Kakatiya Dynasty, which ruled the area from the mid-12th century for about 200 years. The Kakatiya kings developed Warangal as a centre for art, culture and the major part of the epic Sanskrit story 'Mahabharata' was translated into Telugu (the official language of AP) under this Dynasty. Thousand Pillars Temple, Bhadrakali Temple, Ramappa Temple and Warangal Fort are the masterpieces of Kakatiya sculpture and architecture. The dynasty constructed huge tanks in this district to develop agriculture—Ramappa, Pakhal, Ganapa Samudram and Lakhnavaram are the major tanks. One is shown in Figure 4.18.

There has been substantial institutional development of water resources management at the District level over time. This forms the context in which land and water management planning can be developed (Table 4.1).

Hydrological Setting

The district experiences a monsoonal environment similar to much of south-eastern India. The normal annual rainfall is about 1048 mm with 83% of that occurring in 4 months period from June to September. The rainfall decreases from east to west.

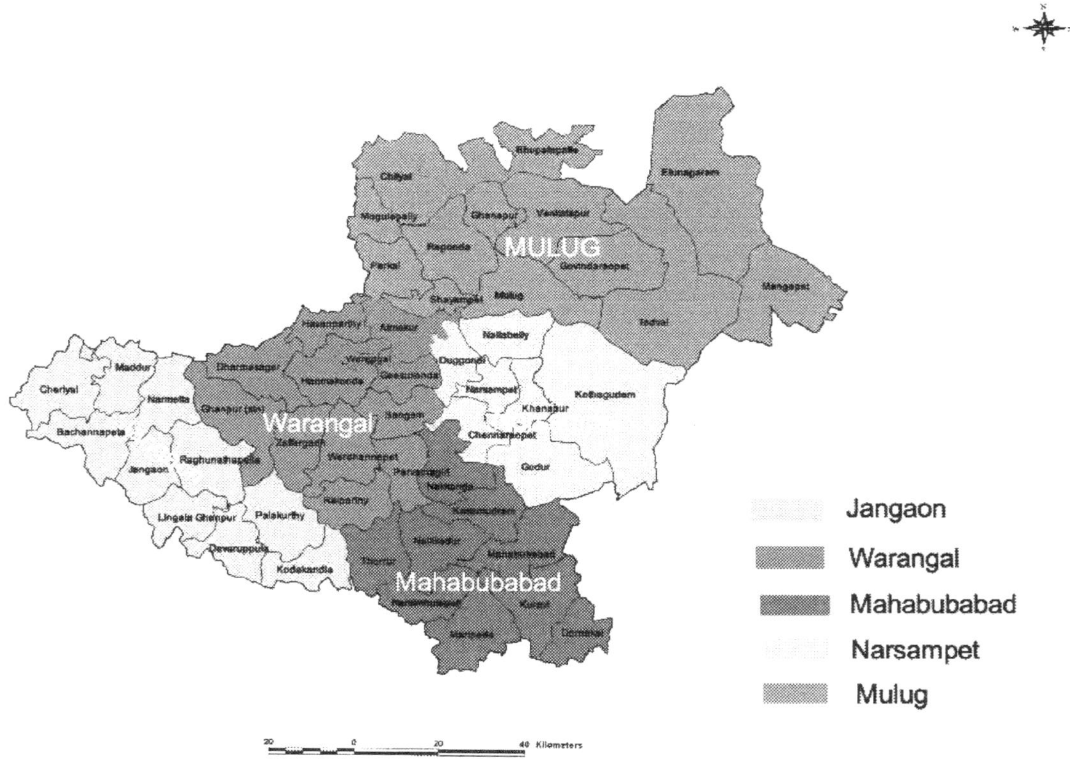

Figure 4.17 Administrative divisions of Andhra Pradesh.

Figure 4.18 Tank used for irrigation in the town of Warangal.
Source: Author.

In the north-eastern areas of the district, the rainfall is about 1200 mm and in the west it is about 750 mm. In 8 out of the last 13 years, the rainfall was much less than average. The mean maximum temperature is about 40°C occurring in April and May and minimum temperature about 13.5°C in December.

The district's total (surface and ground water) water resources are 4.8 bcm out of which 1.4 bcm are currently being used for all purposes. The per capita water resources availability at 1490 m^3 is slightly more than the State's per capita availability of 1400 m^3. The per capita utilisation is about 750 m^3, constituting about 54% of available water. According to international norms, the district can be considered as water scarce and is heading to severe scarcity condition if the present trend of increase in population continues. The western portion of the district with less surface and ground water potential is already facing severe water conditions and the conditions are likely to deteriorate further if remedial measures are not taken.

The district lies in both Krishna and Godavari basins, the major river basins of Andhra Pradesh (see Annex 1 for data & maps). The surface water potential has been estimated by the State's irrigation department by considering the district as two sub-basins of Godavari (G^6 and G^{10}) and three sub-basins of Krishna basin (K^{10}, K^{11} and K^{12}). The total surface potential has been estimated as about 3.0 bcm. The utilisation is more than the availability in the three western sub-basins.

Table 4.1 District Level Committees addressing the water issues

Committee/Board	Chairman	Co/Vice/Ex-Chairman	MS/MC	Members	Roles and functions	Remarks
District Review Committee (DRC)	Nominated Minister	Chairman, ZPP	DC	All MLAs and MPS from district	Review and monitor implementation of all developmental schemes, ensure coordination, recommendations for accelerated development, suggest schemes, monitor utilisation of funds, institute system of addressing public grievances	Planning, GO. DRC assisted by a Technical Group chaired by DC and officials from development departments
District Level Governing Body (DLGB)	Chairman, ZPP	DC	PD (DRDA)	All MLAs and MPs, Officials from Lead Bank, NABARD, M and H, DWMA, FD, HD, AHD, GWD, ITDA, etc.	Coordinate and oversee plans for implementation of SGSY (Village level Self-employment) Scheme, review watershed development programmes, secure inter sectoral and inter departmental coordination	PR and RD; GO Irrigation, RWS and MA and UD are not members
Irrigation Advisory Board (IAB)	DC		SE	All MLAs and MPs, Chairman and representative presidents of WUAs, AD, HD, Revenue, CPO	Irrigation and drainage of the district, canal closure and opening—crop calendar, planning for irrigation and drainage works, environmental issues related to irrigation and drainage	I and CAD, GO RWS, MA and UD, Fisheries, Industry and Environment are not members
District Water and Sanitation Mission (DWSM)	Chairman, ZPP	DC	Nominated by DC	All MLAs and MPs, ZP standing Committee Chairman, CEO of ZP, officials from RWS, Education, M and H, PRO, Social Welfare, Women & Child welfare, DWMA	Planning, coordination and management of RWS project, selection of resource agencies, capacity building, selection of GPs/habitations, ensure district level convergence (water supply, sanitation, health, education, watershed etc.)	PR and RD; GO Supported by District, Mandal, GP and habitation level Water and Sanitation Committees

Continued

Table 4.1 District Level Committees addressing the water issues (*Continued*)

Committee/Board	Chairman	Co/Vice/Ex-Chairman	MS/MC	Members	Roles and functions	Remarks
District APWALTA authority	DC		PD, DWMA	1 MP, 2 MLAs, 3 MP presidents, 2 ZPP territorial constituency members, officials from agriculture, irrigation, RWS, GWD, Mines and Geology, FD, ITDA, APPCB, MA and UD, 5 non-official members	Promote water conservation and tree cover, regulate the exploitation of ground and surface water	PR and RD; Act Mandal level authority
District Water Conservation and Utilisation Committee (WCUC)	Nominated Minister	DC	PD (DWMA)	All MPs, Chairman ZPP, 1 industrialist, 2 experts, 5 WCs/WUAs/VSSs, 3 NGO representatives	Develop a clear vision and strategy for water conservation and sustainable utilisation of water, ensure involvement of the local people, time bound action plan, monitor and oversee the implementation, convergence of programmes of various departments of working in the field of water conservation	PR and RD, GO Mandal, constituency and village level committees
AP Rural Poverty Reduction Project (Velugu)—District level Poverty Eradication and Employment Generation Mission	Nominated Minister		DC	Officials from DWMA, DRDA, APRPRP, SC, ST and BC corporations, etc.	Ensure convergence of poverty eradication programmes, review and advise the respective departments on the programmes, review and advise on the employment generation programmes	

Key: ZPP—Zilla Praja Parishad; GP—Grama Panchayat; DC—District Collector; MLA Member of State Legislative Assembly; MP—Member of Parliament.
Source: Andhra Pradesh Water Vision Task Force (2004, p. 31–32).

The ground water potential of the district is about 1.78 bcm out of which 0.82 bcm is the draft for all uses. The groundwater stage of development for the entire district is 46% with 25% in tank and canal commands and 53% in non-commands. However, about 34% of the district's area are facing over draft of groundwater.

The Warangal District straddles the basin divide of the Krishna and Godavari Rivers. This boundary issue is a critical issue in IRBM. The District LWMP has the opportunity to create workable solutions to local water management within the context of two very large river basin planning procedures which will occur at a higher level of administration. The challenge is to ensure what occurs in the District is congruent with these much higher level and larger initiatives.

Agriculture is the main stay of this rural district, with about 42% of the workers are either cultivators or agricultural labourers. Of the total geographical area of 1.28 million hectares, the net sown area in 2000–2001 was about 0.53 million hectares (41%). The net sown area is very high in the central mandals ranging from 50% to 80% of geographical area of the mandals. About 0.113 million hectares has been sown more than once. The average land holding is 1.32 hectares with 60% holdings less than 1.0 hectares.

Approach to Building a LWMP

The approach used to develop a LWMP for the Warangal District was to use a pressure-state-response (PSR) model. The model is allied to an adaptive management approach and can be used to design best management options within a sub-basin management plan. The PSR model uses indicators to monitor progress made and identifies measures to improve water resources management at smaller scales. The PSR model describes the pressures in a catchment or sub-catchment on environmental themes, the condition or state of each theme and the management responses in the catchment or sub-catchment to mitigate or address known or perceived threats to environmental quality (McAlister et al. 2002). A State of the Environment reporting format is now widely used in many countries and is based on the Organisation for Economic Cooperation and Development's (OECD) PSR model. It was developed in 1991 and has been refined for local application in various parts of the world. The model attributes are illustrated in Figure 4.19 and are described with reference to the conditions existing in the Warangal District. A SoE format can be used to report on progress made in environmental conditions (both natural and human) when implementing the Water Vision in Warangal District.

Environmental indicators can be used to measure condition of the environment in three subregions of the district and the extent to which management actions improve or deteriorate conditions. Environmental indicators are measurable aspects of environmental State and Pressure themes that can be used to report the condition of natural resources, human living conditions, environmental quality and economic conditions. They have two specific functions.

(1) They reduce the number of measures that would normally be required to give an exact representation of a situation.
(2) They simplify the communications process of presenting results to the user.

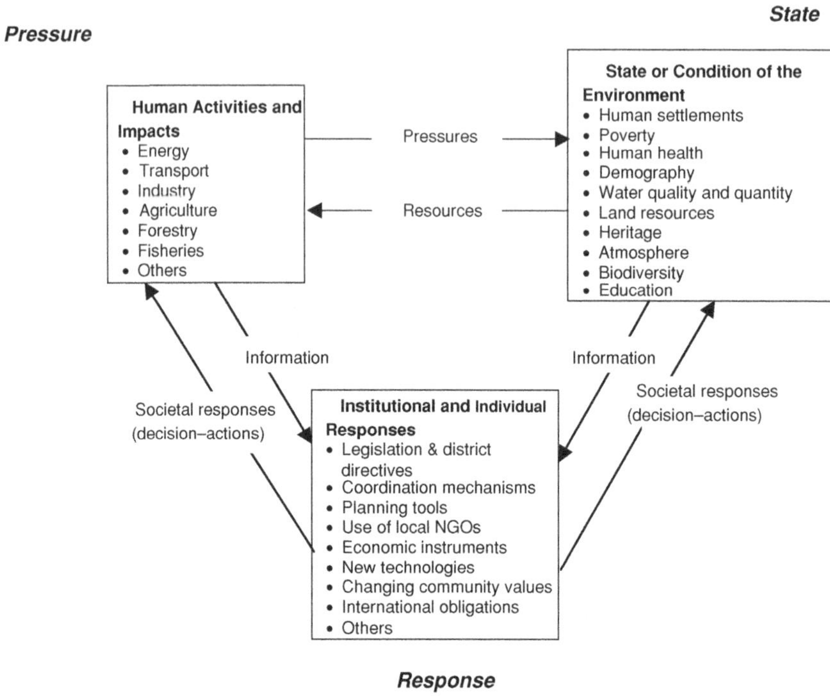

Figure 4.19 Pressure-State-Response model.
Source: Adapted from McAlister et al. (2002).

Pressures are human activities and actions that cause changes in the state of environmental resources. Examples include urban development, polluted industrial discharges and the introduction of exotic species into rivers. Pressure indicators are measures that describe both positive and negative pressures on the environment. *State* refers to specific, measurable aspects of the environment that provide an index of the quality and/or quantity of these resources, and which can be monitored for changes over time. *Responses* are actions by individuals, organisations and governments aimed at addressing adverse environmental changes. Responses may aim to reduce/eliminate pressures or may directly improve condition (McAlister et al. 2002).

Management Options

There is the need for *prescriptive management options* (the responses in the PSR model) which address the specific needs of the Warangal district. The overwhelming need is for the provision of drinking water, both in the rural and urban sectors. Drinking water resources for all habitations needs to be identified, protected and managed in a sustainable manner.

Water management for agriculture both in the rainfed and irrigated sectors needs careful attention. In rainfed agriculture, the requirements of the areas in the low rainfall zone and high rainfall zone (in relative terms) need to be approached separately. Similarly, irrigated agriculture in the canal command areas, under minor irrigation tanks and under tubewells is to be considered in their contexts. Animal husbandry is an important activity and water needs both for livestock and related activities.

Even though industrial development in this District is somewhat limited at present, future water requirements and possible contamination to the water resources from these activities need to be considered. All water quality aspects including wastewater reclamation and recycling will have to be taken into consideration.

Management of water resources is generally based on 'hydrologic unit', principle wherein the hydrological units could be surface water basin, ground water basin or command area of an irrigation canal, etc. These units usually do not follow the administrative units like districts or mandals. In developing the LWMP for Warangal district, while considering the hydrologic units for planning, implementation will normally be at the mandal level as it is there the implementation process usually commences.

Depending on the water scenario, the water development options and demand management options need to be prioritised across the district. The district was divided into *three resource management regions* in the Warangal LWMP (Figure 4.20).

Resource Management Region 1

This management unit is the water scarce *western part* of the district and includes the Janagoan revenue division and western parts of Warangal and Mahbubabad revenue divisions. The two urban areas of Warangal and Janagoan are located in this area.

Pressures

This area experiences widespread degradation of land and water resources and in some locations there is evidence of growing desertification. This is due primarily to the overexploitation of groundwater in recent years due to the prevalence of drought conditions.

State

The hydrologic limitations of the area are

- less annual rainfall, 730–900 mm, with periodic deficit years;
- soils of less water holding capacity;
- surface and ground water potential over utilised;
- to a greater extent, the scope for rain water harvesting is utilised;
- the import of Godavari waters is also limited due to higher topographical elevations and
- excessive fluoride concentrations exist in ground water than are permissible for drinking.

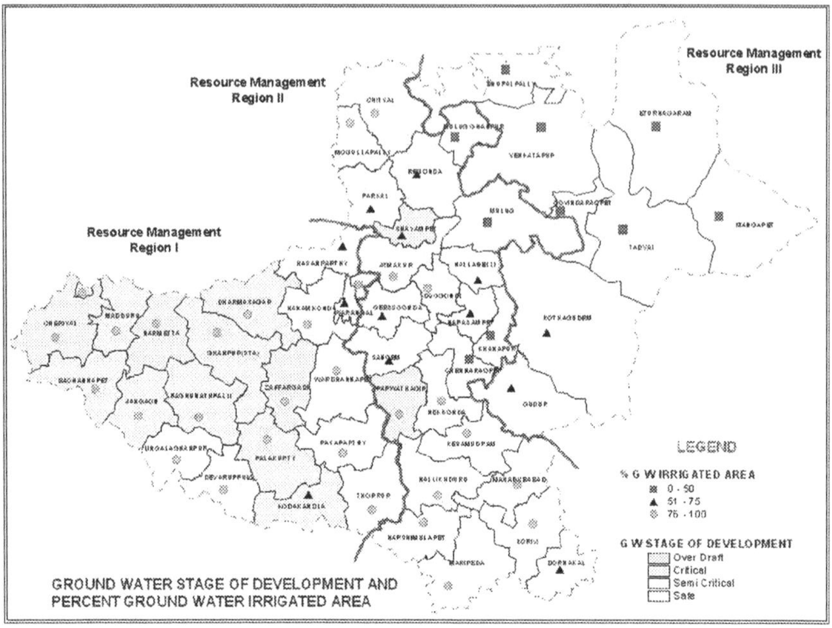

Figure 4.20 Resource management regions of the Warangal district.
Source: Andhra Pradesh Water Conservation Mission.

Responses

The management options for this region were developed in the Water Vision and include:

- Measures to provide secure drinking water supply to both rural and urban areas;
- Identification of areas with fluoride problems and take ameliorative measures;
- Projects to bring in surface water for drinking water needs;
- Protection of existing ground water sources for drinking water needs;
- Extensive adoption of water demand management practices in agriculture with aggressive water conservation practices to free-up water for drinking and water related livelihood activities of poor people, and industrial development;
- Adoption of in situ soil moisture conservation; matching the agricultural activities to land capability; promoting alternative land-use systems and allied on-farm and off-farm activities;
- Strengthening rainfed agriculture; adopt land, soil and crop management practices for efficient use of rainfall including improved agronomic practices;
- Efficient water use in irrigated agriculture; large scale promotion of drip and sprinkler irrigation techniques; shift the focus from increasing productivity of land to increasing the productivity of water;
- Move away from rice and other water demanding crops;
- Effective implementation of Andhra Pradesh Water, Land and Trees Act to check over exploitation of ground water for agriculture;

- Reallocation of water from low-value, typical agriculture uses and inferior crops to high values such as industry and high value agriculture (e.g. horticulture, poultry, animal husbandry);
- An effective link between agricultural production and processing needs to be established in this region; the agro-processing service sector and industry should receive a high priority in order to minimise water demand and benefit larger populations by enhancing the productivity of water;
- Improving the productive use of privately owned and commonly owned non-arable areas through fodder and forage production, establishment of commercial tree crops and dryland horticulture;
- Collection, treatment and reuse of waste water in urban areas (Warangal and Janagoan) for agriculture and other uses and
- Prevention of pollution of surface and ground water.

Resource Management Region 2

This is located in the *central part* of the district and includes parts of Warangal, Narasampet, Mahabubabad and Mulugu revenue divisions.

Pressures

The rich water and soil endowments of the region produces a development pressure. Water is not yet a limiting resource and there is much scope to use water development to benefit poor people. The region should be able to generate a surplus with every drop of water. Providing reliable irrigation services would be the key to improve the irrigation performance in high-potential areas. With reliable service, farmers would invest more in improved technologies and practices, and thus could produce more. With unreliable services, farmers tend to choose strategies that minimise risks, and such strategies are therefore not necessarily profitable or productive. There are ongoing irrigation developments like Kakatiya and SRSP stage II projects, which are planned to bring an additional area of about 175,000 hectares under canal irrigation.

This region has considerably unutilised groundwater potential. Groundwater can be accessible to a larger number of users; it can provide cheap, convenient, individual supplies. It is generally less capital-intensive to develop; but operation costs can be high, depending on the cropping pattern, irrigation techniques and energy charges. Groundwater development is also largely self-financing and is largely private development.

State

This resource management unit is endowed with well-supplied water and soil sources. The favourable hydrological characteristics are:

- average annual rainfall in the range of 900–1100 mm;
- well-known large lakes like Ramappa, Pakala and Lakhnavaram are valuable irrigation sources;

- ground water potential exploitation is in safe to semi-critical range;
- surface water potential is available for development in Muneru and Lower Godavari sub-basins;
- some of the area has black soils, which have a high water holding capacity and
- municipal wastewater management problems are widespread.

Responses

The management options include:

- Improved management of ground water both by using water saving methods like drip and sprinkler methods as well as sustaining the ground water resources by recharge of aquifers using available rainfall;
- Improving drinking water quality and quantity;
- Effective implementation of APWALTA to check over exploitation of ground water for agriculture and safe guard drinking water supplies;
- Rehabilitation of the tanks through desilting, enhancing the inflows and reducing silt inflows into tanks by undertaking soil and water conservation measures in the catchment areas;
- Improved tank management with appropriate allocations for irrigation, livestock use and aquaculture; many of the tanks may not sustain aquaculture through out the year, but with appropriate management the length of water availability for fish production can be increased; small water bodies can be used to raise fingerlings;
- Adoption of modern agriculture with high land and water productivity; productivity gains would be achieved from improved agricultural practices and improved water delivery services;
- In agriculture, high-value enterprises (such as spices, horticulture, fisheries, etc.) need to be encouraged to achieve a higher growth and
- The available potential needs can be harnessed using a conjunctive planning approach.

Resource Management Region 3

This management unit is the *eastern part* of the district and comprises of some mandals from Mahbubabad, Narasampet and Mulugu revenue divisions, which have a high percentage of forest area.

Pressures

The region experiences fewer pressures than Regions 1 and 2. There is the need to address poverty by developing intensive agriculture in already converted agriculture lands with adequate water development for irrigation to benefit poor tribal people. Other pressures include unauthorised land-use conversion from forest to agriculture, poor infrastructure and market facilities, inadequate medical and health care and education facilities.

State

The area receives more than 1000 mm rainfall annually and has very high ground water and surface water potential. No water quality problems exist and forest based livelihoods and industries are possible.

Responses

The management options include:

- Land management practices for the effective use of water, especially in lands recently converted to agricultural use;
- Improving drinking water quality and quantity;
- Improving animal husbandry to add incomes to agricultural income;
- Productive use of presently non-arable areas through fodder and forage production, establishment of commercial tree crops including horticultural plantations;
- Improved forest management through Joint Forest Management (Vana Samrakshana Samithi (VSS)) to enhance forest cover and to provide usufructuary benefits to the members and
- Improve aquaculture in tanks, rainwater harvesting structures and the Godavari river.

Draft Contents

At the time of writing, draft Warangal LWMP contents had been developed (Table 4.2). The Plan guides preferred management options relevant to the most critical resource management issues for the three resource management regions and at mandal level.

Institutional Arrangements for Improved Coordination

The Warangal LWMP will be implemented using several institutional arrangements: the organisations for district water management, coordination mechanisms between agencies, and harnessing NGOs, academic institutions and other water stakeholders to work together; and funding mechanisms for sharing the costs of land and water management.

Water is basically a state subject, and so is its development, utilisation and monitoring. The Government of Andhra Pradesh is responsible for water resources planning, and the storage and use of its water resources. The water resources of the inter-state rivers are governed by water allocations between states. As the Warangal District is part of basins of two inter-state rivers, water developments in the district have to consider inter-state water sharing. This is yet to be done. Several Government departments, agencies and people's institutions are involved in water development, use, monitoring and regulation. The state-level institutional framework and the legislative framework are described in AP Water Vision Volume I document (Andhra Pradesh Water Conservation Mission 2003).

There is a proposal to constitute *District Planning Committees* through an enactment to plan, coordinate and monitor all developmental activities including

Table 4.2 Warangal Land and Water Management Plan—Draft table of contents

Chapter	Title	Contents
1	Introduction	(a) AP Water Vision and district issues (b) Purpose of the document (c) Need for IWRM (d) What is IWRM, elements, functions (e) Engagement/participation methods used in plan developed specified
2	Status of natural and human resources	(a) Geographic information systems to be developed based on current WCM Warangal District GIS (b) Land systems approach used and land management units with best management options and photos of each LMU included in GIS (c) Method to be developed to link GIS to district decision-making processes (d) Health and poverty indicators, and other selected data layers to be included in GIs, focusing on water resources management issues and water data. (e) State of environment (indicator) layers included in GIS
3	Water Demands and Issues	(a) Specification of both—using data in the Water Vision document
4	Current institutional arrangements for land and water management	(a) Description of current arrangements (b) SWOT analysis used to identify gaps, including current and needed coordination mechanisms in the District Collectors Office
5	Institutional framework to implement AP Water Vision at District level	Preferred arrangements will include: (a) Plan making process in District Development Plans for the three resource management regions (b) Specification of work of the District Collector and agency staff at Warangal, including coordination and reporting mechanisms
6	Implementation plan	(a) Specify roles and responsibilities of government agencies, NGOs, peoples' institutions and others (b) Coordination mechanisms within the district between agencies, NGOs, District Collector, and other institutions (c) Monitoring and evaluation processes (d) Reporting and accountability mechanisms (e) Mechanisms for improved management and enforcing environmental laws
7	Monitoring	(a) Specification of monitoring methods and indicators to be used to measure improvements made from implementing LWMP Plan
8	Research needs	(a) Identification of knowledge gaps (b) Development of research priorities and a Warangal District research plan (c) Development of a catchment information management system to retain current knowledge on best management options for each resource management unit

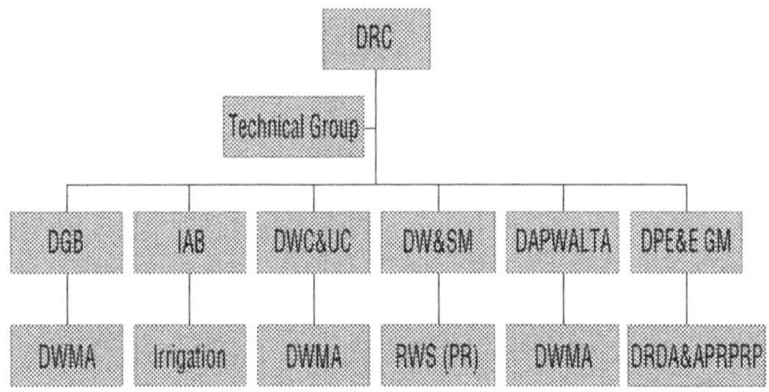

Figure 4.21 Current Warangal District Panchayat Raj Structure.
Key: Technical Group: This group consists of Heads of Technical Departments and Agencies who plan, execute and monitor the progress of their respective programs like agriculture, irrigation, rural water supply, animal husbandry, health, etc.
Source: Andhra Pradesh Water Conservation Mission.

natural resources related activities. At present a *District Review Committee* headed by a Minister is monitoring the activities. A brief sketch of the current district institutions includes:

- The *District Collector*, who is the key position heading the administration and coordinating all the activities. The Zilla Praja Parishat[3] Chairman is the head of the district level Panchayt Raj institution. Although all the water departments have their presence at district level (Figure 4.21), their structures are different.
- The *Irrigation Department* has presence in the district at the level of Superintending Engineer and supported by three divisions in the Warangal district. They look after the minor and medium irrigation in the district. The major irrigation projects, which occupy more than one district, have separate project chief engineers and divisions. In Warangal district, the project staff of Kakatiya canal, the Sri Ram Sagar Project (SRSP) Stage II, SRSP Flood Flow canal and Godavari Lift Scheme are also located. They are mainly engaged in the construction of the irrigation networks and command area development.
- The *Panchayat Raj institutions* covering minor irrigation (village ponds) and rural water supply are at district and Mandal levels. Watershed development is headed by Project Director (District Water Management Agency) with three multi-disciplinary teams comprising engineering, agriculture and animal husbandry disciplines.
- The *water user departments* such as agriculture, horticulture, animal husbandry and fisheries, rural water supply, industries have district level offices; only agriculture and rural water supply have mandal level offices. Public health and Municipal Engineering offices are located at the two urban centres of Warangal and Janagoan.

[3] The Zilla Praja Parishat is the elected body at district level.

- The *ground water department*, which monitors ground water conditions, is headed by the Deputy Director. The AP Pollution Control Board has a regional office in Warangal. The Medical and Health Department is headed by a Deputy Director.
- The regional centre of *Warangal* is a well-developed *educational centre* with the National Institute of Technology, Kakatiya University, Kakatiya Medical College, Kakatiya Institute of Technology and Sciences offering education in engineering, medicine and other major disciplines. Research facilities include the National Institute of Technology, Regional Agricultural Research Station of Acharya, N.G. Ranga Agricultural University at Warangal and a Horticultural Research Station of the Agricultural University at Malyala of Mahbubabad Mandal. There is also the Krishi Vigyan Kendras DAATTC (District Agriculture Advisory and Transfer of Technology Centre) for disseminating agriculture technologies.
- *Several water development and user departments* have district level advisory and monitoring bodies (Table 4.1; Figure 4.20). These bodies are headed by a Minister or ZPP Chairman or Collector with elected Members of Parliament, Members of State Assembly and of local bodies, representatives of People's Institutions and the departmental officials as members.
- *People's Institutions*. A close partnership between the people and the Government is being promoted in order to achieve sustainable use of water resources. There are a large number of such institutions in Warangal district. In the 51 mandals in the district, there are 682 water user associations, 126 watershed associations, 218 VSSs, 1010 Functional Committees for Natural Resource Management and Water Conservation and Utilisation Committees. The Rural Water Supply and Water Conservation and Utilisation Committees are at each village, mandal and district level. The Water Use Associations are based on irrigation command area basis. Watershed Associations and VSSs are project-based bodies. In other words, there exists a highly developed grass-roots natural resources management process.

The Warangal LWMP sets up a process of coordination performance appraisal of local natural resources management. This will be done through restructuring of the local Panchayat Raj Institution to create a District Water Mission which will report to the District Collector (Figure 4.22).

The roles and responsibilities of this committee are to

- ensure coordination between government agencies to deliver water resources management service in the District;
- establish a monitoring and reporting procedure to evaluate its success in implementing Water Vision objects for each of three Resource Management Regions in the District;
- provide a District database and GIS, in liaison with the Water Conservation Mission to record land-use changes, environmental performance improvements and other State of the Environment information and

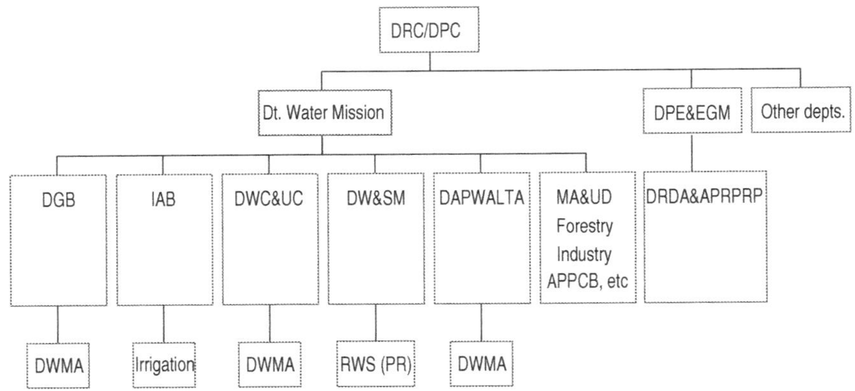

Figure 4.22 Membership and reporting lines of the proposed Warangal District Water Committee.

Key:
DRC/DPC: District Review Committee/District Planning Committee
DPE & EGM: District Poverty Eradication & Employment Generation Mission
DGB: District level Governing Body
IAB: Irrigation Advisory Board
DWC&UC: District Water Conservation and Utilization Committees
DAPWALTA: District Andhra Pradesh Water and Land an Trees Authority
APPCB: Andhra Pradesh Pollution Control Board
MA&UD: Municipal Administration & Urban Development
DRDA: District Rural Development Agency
APRPRP: Andhra Pradesh Rural Poverty Reduction Project
DWMA: District Water Management Agency
RWS (PR): Rural Water Supply (Panchayat Raj Department)
Source: Andhra Pradesh Water Conservation Mission.

- be responsible for enacting water components of the proposed changes to the role of the District Collector, under the proposed A.P. District Planning Committees Act. It is expected that when in place the District Collector will have strong local authority in water matters and will report directly to the Chief Minister of the State.

Implications of the case for other areas

The future of the Warangal LWMP lies in the ability of State Government to enact local coordination of natural resources management. The proposed structural changes to district administration, and increased implementation of projects and monitoring will be a slow process. If the mechanisms established in this process are carefully designed and implemented, there is the opportunity for success. The most recent workshops (2004) in the district suggest that local leadership, interest from local research institutions and State government support are strong, but it will require top–down enabling: by the passing of the proposed A.P. District Planning Committees Act and by head office agencies delivering funding support to the process.

The Warangal District LWMP is a useful example of how IRBM can be delivered at the local level. Here, the District is the local enactor of a hierarchical approach as outlined in Figure 4.7. There are several characteristics of this case study which suggest the approach to sub-basin land and water resources management can be applied elsewhere:

- High population densities creating urgent demand on water resources and the need for developing immediately accessible, affordable, continually supplied, clean potable water supplies.
- Varying hydrological conditions, including highly stochastic rainfall regimes (often with monsoon failure or lateness) which require a differential spatial specification of efforts (hence the three sub-region approach).
- An emerging administrative model which favours local, district action in natural resources governance.
- Evidence of strong local ownership of resource management issues.
- Well-developed and functioning local educational and research infrastructure, water user associations and peoples' involvement groups.

These characteristics are common in other parts of south-eastern India and, to some degree, in other parts of monsoonal and sub-monsoonal Asia and south-east Asia. There is the opportunity to transplant the LWMP approach to such regions, but it is recommended that well developed, thorough water visioning is undertaken before the start of a LWMP. This will allow local involvement and ownership, articulation and specification of issues, clarification of the roles and responsibilities of district level governance (of the District Collector, State agencies, peoples' institutions, NGOs) and development of coordination mechanisms. The water visioning will include a strategic statewide water framework which specifies the state of the land and water environment, the pressures on those resources, the management options to improve conditions, and a monitoring procedure to measure gains and losses.

The limitations of this case study for use elsewhere include a lack of IRBM planning which occurs at a higher level and within which this approach could be embedded. However, there are proposed river basin scale water resources management plans in the State and it is axiomatic that these should recognise the significant achievements already made in the Andhra Pradesh Water Vision and the Draft Warangal LWMP.

Second, there is the problem that this case exists in a water policy 'vacuum' in the State of Andhra Pradesh. Only rudimentary water policy exists and limited new policy development has occurred. Until further policy formulation is completed, district plans will suffer from the lack of a clear enabling environment and institutional mechanisms to drive action (policies, laws and financing mechanisms) to support local efforts.

Third, the project which supported the Draft Warangal LWMP and the Andhra Pradesh Water Vision was funded externally (Royal Netherlands Embassy) and the work has come to an end. There is now a funding gap and the need exists by the State water institutions to own and run the process. There was some evidence that local support will materialise (at the time of writing, 2004).

Fourth, the project has raised expectations of successful outcomes. Without ongoing commitment by State agencies, there is the danger that achievement will not be realised in the short to medium term (2–5 years) and disappointment and disinterest will set in. Like other experiences documented in this book (Chapter 2), realistic and achievable short-term gains must be made to stimulate ongoing support.

Fifth, there is the danger that the process relies on knowledge external to the region to drive process. What is needed is immediate training and equipping of water agency professional staff, NGO officers, field extension staff, leaders of water user associations and other water and development groups in methods of group decision-making, coordinated planning and management, monitoring, mentoring and the overall concept of integrated water resources management. Much could be done by simply enforcing water laws in the district and improving village level potable water supplies, coupled with training in coordinated management across agencies.

Despite these challenges, this case offers a template for local action in IRBM. The challenge is to learn from this experience as it evolves and translates the learnings elsewhere. One option is to test this approach (district LWMPs) in other parts of Andhra Pradesh before it is applied to other Indian states or countries, learn from these experiences and modify the approach to make it more appropriate to other local conditions. This knowledge can then be presented as guidelines for effective LWMPs.

4.4 SUMMARY

In this chapter, we have seen how decision-making for IRBM is a complex and adaptive process. A protocol for IRBM was presented, designed to respond to this environment. National and international RBM protocols were presented and a case study of district level land and water management planning (a sub-basin, local initiative) demonstrated how IRBM actions can be developed at local levels.

The challenge for IRBM protocols in any location, whether in national or international basins, or at the local level, is to ensure that natural resources management is well-informed. IRBM is informed by dialogue, and basin managers need to get the basin knowledge and wisdom correct if effective decisions are to be made. This issue is developed in the next chapter.

4.5 REFERENCES

Andhra Pradesh Water Conservation Mission. 2003, AP Water Vision. A Secure Water Future. Vol. 2. Hyderabad, Water Conservation Mission, Government of Andhra Pradesh.

Andhra Pradesh Water Vision Task Force. 2004, Land and Water Management Plan. Warangal District (Draft). Hyderabad, Mission Support Unit, Water Conservation Mission.

Chave, P. 2001, The EU Water Framework Directive: An Introduction. London, IWA Publishing.

Christian, C. S. S. G. A. 1968, Methodology of Integrated Surveys. Aerial Surveys and Integrated Studies. Proceedings of the Toulouse Conference, UNESCO. 1964. 233–280.

Cocks, K. D., Ive, J. R., Davis, J. R., & Baird, I. A. 1983, SIRO-PLAN and LUPLAN: An Australian approach to land use planning 1. Environment and Planning B 10, 331–345.

Comino, M. P. 2003, The Healthy Rivers Commission: A New Approach to Integrated Governance. Healthy Rivers Commission, Sydney.

Council of the European Union. 1995, Helsinki convention: Trans-boundary watercourses and international lakes. Official Journal L 186.

Ewing, S. A., Grayson, R. B., & Argent, R. M. 2000, Science, citizens, and catchments: Decision support for catchment planning in Australia. Society and Natural Resources 13[5], 443–459.

Global Water Partnership. 2002, Integrated Water Resources Management ToolBox. Stockholm, Global Water Partnership.

Golub, J. 1996, Why did they Sign? Explaining EC Environmental Policy Bargaining. San Domenico: Italy, European University Institute. EUI Working Paper RSC No. 96/52.

Government of India. 2000, National Water Policy. Publisher: Government of India, New Delhi.

Gregg, F., Born, S. B., Lord, W. B., & Waterstone, M. 1991, Institutional Response to a Changing Water Policy Environment. Tucson, AZ, Water Resources Research Center, University of Arizona.

Healthy Rivers Commission of New South Wales. 2003, Healthy Rivers for Tomorrow. Sydney, Healthy Rivers Commission of New South Wales.

Hooper, B. P. 1995, A Scoping Paper on Water Quality Management in Malpas Dam and its Catchment. A Report to Armidale City Council. University of New England, Armidale, Centre for Water Policy Research.

Hooper, B. P. 1997, An integrated catchment management protocol for water quality management in Australia. Water Quality International Jan/Feb, 48–50.

Hooper, B. P. 1998, Elements of Best Practice in Managing Catchments for Water Quality. Paper prepared for the Sydney Water Inquiry Secretariat. Brisbane: Integrated Resource Management Research Pty. Ltd., 1–4.

Hooper, B. P. & Duggin, J. 1996, An ecological approach to floodplain zoning. Land Use Policy 13[2], 87–99.
Hooper, B. P. & Lant, C. (in press), Integrated, adaptive watershed management. Hanna, K. S. & Slocombe, S. D. (eds.). Fostering Integration: Concepts and Practice in Resource and Environmental Management. Oxford University Press, Oxford.
House of Representatives Standing Committee on Environment and Heritage. 2000, Co-ordinating Catchment Management: Report of the Inquiry into Catchment Management. Canberra, Australia, Commonwealth of Australia.
Ive, J. R. & Cocks, K. D. 1983, SIRO-PLAN and LUPLAN: An Australian approach to land use planning 2. Environment and Planning B 10, 347–355.
McAlister, T., Richardson, D., Thomson, D., Veary, A., Agnew, L., & Burgess, R. 2002, Lake Samsonvale Integrated Catchment Management Strategy. R.B 13088.005.00.ICMS.doc. Brisbane, WBM Oceanics.
Millington, P. 1999, River Basin Management—Its Role in Major Water Infrastructure Projects (Draft). Cape Town, South Africa, World Commission on Dams Secretariat. World Commission on Dams Thematic Reviews. Institutional Processes V.3 River Basins—Institutional Frameworks and Management Options.
Millington, P. 2004, Integrated River Basin Management—From Concepts to Good Practice, Briefing Notes to the World Bank (Draft).
Mitchell, C. 1991, Terrain Evaluation. An Introductory Handbook to the History, Principles and Methods of Practical Terrain Assessment. Longman Scientific and Technical London.
Mostert, E., Van Beek, E., Bouman, N. W. M., Hey, E., Savenije, H. H. G., & Thissen, W. A. H. 1999, River Basin Management and Planning. Keynote paper for International Workshop on River Basin Management. The Hague, International Workshop on River Basin Management.
Newson, M. 1992, Land, Water and Development. River Basin Systems and their Sustainable Management. New York, USA, Routledge.
Shah, T., Molden, D., & Sakthivadivel, R. 2003, Limits to Leapfrogging: Issues in Transposing Successful River Basin Institutions in the Developing World. Colombo, Sri Lanka, IWMI-TATA Water Policy Program. Water Policy Briefing.
Walters, C. J. 1986, Adaptive Management of Renewable Resources. New York, NY, USA, Macmillan Publishing Co.

5

Information systems for IRBM

5.1 THE RIVER BASIN MANAGEMENT INFORMATION PROBLEM

5.1.1 Recurring problems, few solutions

Effective IRBM requires accurate, up to date, relevant information. There are increasing concerns about handling the growing amount of information emanating from research projects, community sources, government databases, reports and projects about natural resources management and applying it to the decision-making processes in IRBM. These information sets and data comprise a large, complex array of biophysical, social and economic information coming in both qualitative and quantitative forms.

These concerns are not new. In 1959, a meeting was held in Knoxville, Tennessee, the home of the Tennessee Valley Authority, one of America's oldest and internationally acclaimed river basin management organisations (RBOs). The purpose of the meeting is found in the final report (Tolley and Riggs 1961):

> to contribute to a variety of working-level planning decisions that concern technicians, administrators, and legislators in on-going water resources development, focusing on how small watershed development can best fit into river basin and regional planning.

The report developed procedures for meshing watershed management with river basin development, interdisciplinary teamwork in watershed planning and economically desirable institutional arrangements and cost-sharing requirements.

 ISBN: 1 84339 088 4.

These procedures are still being worked out in many places some 40 years later, including in the USA. The report stated that there was a need to develop

> information activities which include technical training, development of effective media to communicate ideas, and an understanding of the issues that have a bearing on the interests of those participating in the watershed program
>
> (Tolley and Riggs 1961)

These issues also remain critical today.

The information needed for effective decision-making is often inaccessible to decision-makers, is hard to organise and changes frequently. The problem is essentially a management problem and is experienced by RBOs, which appear unable to handle the growing amount of information because they lack the human, financial and technical resources to integrate the information to enable local and regional decision-making in natural resources management.

This information load is thought to impede the resource allocation and planning decisions of RBOs, and to preclude them from providing information to their stakeholders about appropriate landuse options. Not knowing how to advise people in a river basin puts an RBO at a comparative disadvantage when land and water use decisions are being made by governments and the private sectors. Yet, it is frequently the role of the RBO to be the ‘advocate’ for river basin management.

The core challenge is to address the fragmentation of the decision-making process and the variety of information outputs required for a large and varied client base. Within this context, there are calls for decision-making to be based on comprehensive assessments of development and management alternatives which take into account biophysical, environmental and socio-economic factors as well as the time and space dimensions of sustainable land management (Lee et al. 1992; Reibsame et al. 1994; Sombroek and Antoine 1994; Shaw et al. 1996).

The river basin management information problem is growing and not new. Past approaches include the use of decision support systems (DSS), geographic information systems (GIS) and databases. Advances in computer technology, particularly in GIS and a corresponding capacity to develop DSS, have revolutionised the way some aspects of resource management decision-making is conducted. DSS are usually (but not necessarily) computerised systems that integrate expert knowledge and models with timely information to assist in day-to-day operational and long-range strategic decision-making (Shaw 2003). DSS for natural resource management have concentrated on facilitating decision-making in relation to one or more issues such as the control of water pollution, ground water contamination or erosion, and for broader watershed management requirements (Tecle 1992; Meijerink et al. 1993; Prato, Fulcher and Feng 1995; Jamieson and Fredra 1996; Reitsma 1996; Walker and Johnson 1996). Laut and Taplin (1989) observe that DSS have been highly tuned to the environment in which they were built. Consequently, while the general form is transportable, considerable adaptation and experimentation of the model is required for satisfactory performance in a different area.

This chapter discusses the types of information systems, concerns about their use and the idea of an integrated information system. The need for stakeholder-driven information exchange is outlined and the principles of information exchange are described. Finally, the dimensions of delivering research information to river basin management are described.

5.2 TYPES OF INFORMATION SYSTEMS

5.2.1 From data to wisdom

What is river basin management information? It is knowledge about the management of different land types within a whole river basin context. This knowledge comes from research projects, modelling, government reports, local government planning instruments, policy documents, statistical surveys, conferences, laws and practical experience of land and water management.

River basin information comes in many forms and styles, for example, written reports, biophysical, economic and social data, management options, and first-hand, verbal experiences. Exchange and building of IRBM information can be developed through an information exchange process, which allows professionals, practitioners and the general public to exchange experiences in implementing IRBM. In this way, exchange and building IWRM information is a capacity building tool.

This involves getting diverse information into the hands of water professionals, especially those in government decision-making agencies, and helping them share information, ideas and experiences. The process places importance on all relevant information sources, not just those from 'technical experts'. It engages local communities by providing opportunities for discussion and simplifies their ability to provide and access information.

There are four broad types of information:

- *data*: quantifiable and qualitative facts about the characteristics of water resources (quality, volumes, location, frequency of occurrence, spatial variability);
- *information*: how these data can be assembled into meaningful patterns for specific purposes;
- *knowledge*: understanding of the implications of trends and values in data over time, personal and corporate understanding of resource use practices and their impacts;
- *wisdom*: agreement about commonly accepted methods of using water resources to ensure sustainability.

In any river basin, this broad array of information is available and the challenge for river basin managers is to choose what to use, assess its reliability and determine how to use it in decision-making. The focus is frequently on the use of innovative or improved management practices and how their effectiveness, if adopted, changes the quality of a river basin's natural resources. This quality change will be recorded in audits on the condition and trend of the natural resource conditions that are often taken over long periods of time.

The development of river basin information systems should focus on addressing needs. The primary need is for appropriate information management systems (IMS). Laut and Taplin (1989) maintained that GIS when properly matched to resource databases and appropriate models and algorithms, hold considerable promise for river basin managers. If backed by adequate display tools, they demonstrate quickly and clearly the severity and location of issues and problems and so enhance public awareness and can be used to overcome public prejudice, ignorance and incipient hostility. If the models and algorithms of a GIS are used with effective decision assistance systems, complex and original investigations of policy options can take place without extensive reprogramming. The combined tools can quickly and graphically demonstrate the likely consequences of policy implementations. The development of policy options for river basin management requires integrated regional models that identify drivers of landscape and landuse change. Once identified, these drivers can assist the evaluation of policies and programs to enhance the adoption of sustainable land and water management practices in river basins.

Critical knowledge gaps often exist in river basin management, and there is the need to identify the gaps as a matter of urgency. These gaps include tools to analyse landuse change and the environmental and social impacts of river basin management plans, which require landuse modifications.

5.2.2 Information management systems

Assisting decision-making

An information management system in its simplest form can be a manual of good practices, shared amongst practitioners. 'Information management system' today is a generic term to describe a range of electronic systems to arrange, store and exchange data and information. Advances in computer technology, particularly GIS and the corresponding capacity to develop DSS have revolutionised the way some aspects of river basin management decision-making is conducted. DSS are usually, but not necessarily, computerised systems that integrate expert knowledge and models with timely information to assist in day-to-day operational and long-range strategic decision-making. To date, however, DSS for natural resource management have concentrated on facilitating decision-making in relation to one particular operation or issue such as the control of water pollution, ground water contamination or erosion (Jamieson and Fredra 1996; Reitsma 1996). As a result, DSS have only been able to offer partial support in resource management decision-making. Furthermore, Laut and Taplin (1989) observe that DSS have been highly tuned to the environment in which they were built. While the general form is transportable, considerable adaptation and experimentation of the model is required for a satisfactory performance in a different area.

In the management of natural resources, we saw in Chapter 2 that people respond to cues from both the physical environment (e.g. soil erosion, salinity) and from the socio-cultural context (e.g. conservation policy) and behave to increase both economic and socio-cultural well-being. IRBM is considerably more complex than the capabilities of existing DSS which address uni-dimensional issues. Complexity is derived from two factors: the context in which resource management is carried out and the inadequacies of the existing resource management decision-making process.

There is a limited but growing number of examples of IMS specifically designed for river basin managers, despite increasing attention being paid to the use of simple, complex and integrated modelling technologies by natural resource management agencies. Despite a clear need to integrate socio-economic and cultural information, there are few empirical models that reach this level of sophistication. Little attention has been placed on defining a suitable framework for the construction of an information management system for river basin managers. There are two distinct elements that need to be addressed in the development of these systems: process issues and modelling issues.

Process issues

Laut and Taplin (1989) proposed that IMS for catchment managers should be based on an understanding of and be linked to current institutional arrangements for regional decision-making. Based on evidence from an information management system developed to manage grazing lands in China, Hamilton and Sheehy (1993) suggest that ideally the implementation of the system should be an institution or organisation that has power in resource decision-making. Agencies that lack such power are likely to be ineffective. The authors suggest that the institution or organisation should possess the following characteristics:

- influence in planning and decision-making with regard to resource management;
- access to and ability to process and use resource information;
- access to internal and external funding sources;
- sufficient prestige and power to bring together all the stakeholders necessary for implementing the information management system;
- long term commitment to the development of an information management system to update and expand databases as new or better information becomes available;
- the capability of establishing linkages with other institutions to provide support for the introduction and implementation of the information management system.

Rapid development in information technology in the last decade has created opportunities for the development of multipurpose information systems that can be used to quickly and efficiently generate information for a variety of users (Sombroek and Antoine 1994). Although, a computer format is not necessary for the development of IMS, the widespread availability of personal computers and the capability to process large amounts of information make computer-based procedures appealing. However, computer-based IMS require a user-friendly interface and adequate graphical capability in order to demonstrate quickly and clearly the implications of a given alternative in a manner easily understood by the end-user. Only then can computer-based procedures overcome attitudes of prejudice, ignorance and distrust towards computers that extension research has identified as prevalent amongst rural communities (Auscher 1993).

Choosing from a range of technologies

Fredra (1996) identified several technologies which lend themselves to the development of IMS: geographical information systems (GIS), databases, expert systems and simulation and optimisation models. These are discussed later in this chapter.

GIS capture, manipulate, process and display spatial or geo-referenced data. The obvious advantage of GIS is that it shows data queries, analyses and evaluations in a geographically referenced format which aids the understanding and interpretation of information. While there is some criticism that there is a disparity between the current capabilities of GIS and the requirements of the end users, many of these problems can be overcome by integration with other technologies. The usefulness of databases in the collection and storage of information is generally recognised. The integration of databases and models allow users to automatically retrieve and load input data in order to assess the consequences of complex ecological and environmental processes (Fredra 1996).

Where more complex decision models are required, expert systems can play an important role. Expert systems rely upon the existence of a body of commonly agreed knowledge, often derived from experimental and empirical evidence which can be systematically linked together to predict outcomes (Laut and Taplin 1989). While expert systems are commonly used for on-going management problems, the possible uses of expert systems in river basin management include the development of policies and management plans for irrigation districts with rising water tables, tree-planting strategies and optimal landuse/vegetative cover for highland river basins. The coupling of optimisation and simulation models offers another way of examining the characteristics and behaviour of certain problems. Optimisation requires simplification of the problem. For example, steady state and/or spatially aggregated descriptions of a particular problem may be used to simplify complex situations. Simulation models are then used to develop more detailed information about the problem, e.g. dynamic and spatial distributed information would be generated (Fredra 1996).

The next sections of this chapter summarise existing technologies. There are four broad groups of activities: meta-data sets, bibliographic and knowledge management systems, GIS and modelling and DSS.

*Meta-data sets**

These data sets are large provincial, national and international sets of data about the earth's resources and resource use. They are frequently housed in government agencies, such as the US Geological Survey. While the data sets vary in type, nature and function, the focus tends to be on national data sets which show resource condition and trend.

New techniques have been developed to include stakeholders' needs in meta-data management for specific purposes, e.g. watershed management, water resources planning, social and biophysical impact assessment. These include interactive group methods, Delphi techniques and microcomputer-enhanced decision-making techniques which use community and expert advice to construct appropriate data management tools.

The use of meta-data is varied, according to macro-scale resource management planning needs. The GWP Toolbox suggests a number of lessons that have been learned from using meta-data (www.gwpforum.org). Meta-data sharing should be:

- based on people management (empowerment and skilling of organisations) in addition to being technology driven;
- demand-driven whereby the needs of the resource managers are clearly articulated at the outset of system design and construction and the outputs are directed toward these end users;
- flexible enough so that the sharing system can be transported and implemented in a variety of locations;
- transparent and rigorous in the way it processes data and information so that technical and non-technical persons can follow the process of alternative generation and evaluation;
- interactive, one which facilitates a participatory decision-making process;
- facilitating, easing the process of learning and increase awareness of the various management and development options available;
- built with the capacity to integrate multidisciplinary information;
- readily accessible to the large number of stakeholders and participants in the resource decision-making process.

Two specific cases of meta-data are:

- The National Land and Water Audit (Australia) developed several measures of resource condition and trend and recommended protocols to exchange state information nationally between often competing jurisdictions (www.lwa.gov.au).

---

*The sections on meta data, information management systems, geographic information systems, modelling and decision support systems draw on the author's contributions to the GWP IWRM Toolbox: www.gwpforum.org

- The Colorado Decision Support System is a state-run integrated dataset broken into various watersheds. Incorporated within the extensive database is a GIS system providing spatial representation of various property rights and land-use activities. The system's availability on-line makes it easy to utilise. While the state maintains ownership of the database, it is in the public domain and is open access.

Bibliographic/knowledge management systems

There are two types of information management systems of particular importance to river basin management: bibliographic information (and knowledge) systems and geographic information systems (GIS). These provide support for decision-making, but are distinguished from DSS.

These include various knowledge management tools, often located within government agencies and commercial companies, to measure the effectiveness of communication processes, including:

- tools to monitor and evaluate the communication system(s);
- mechanisms to ensure correct signals are delivered/received through information exchange processes;
- tools for benchmarking organisational activities to establish best management practices.

Information management systems (IMS) work better when there is knowledge of the social system of a particular setting, to ensure the structure, functioning and applicability of the IMS is relevant to the task at hand in IWRM. IMS work best when a suitable implementation, monitoring and evaluation system is built prior to implementation.

It is important to understand the social structures, gender issues, stage of economic development, human and technological resources and managerial capacities of water resources managers in the settings for which the IMS is designed. IMS can be misused to drive implementation processes, when they should monitor the effectiveness of the result of a specific communication program.

An IMS is only as good as the results it records and monitors. The following indicators can be used to gauge the effectiveness of an IMS:

- the number, type and variety of implementation actions (e.g. farming practices, improved water management plans);
- the number of uses, the frequency type and duration of use of the IMS;
- the degree to which the stakeholders form a coherent social entity to facilitate dialogue and information exchange;
- a range of organisational performance indicators (customer–client dialogue, reporting mechanisms, financial stability and structural change stability of organisations);

- more consistent decisions;
- fewer instances of public conflict as a result of using the IMS by providing specific data on different management options, their acceptability, stakeholder priority and cost;
- reduced resource degradation by implementation of best management options developed in the IMS.

GIS

A GIS is an inventory of natural resources (soil, landform, water and vegetation) of an area of the earth's surface. It is a tool to aid decision-makers with data and information with which they can make water planning and management judgements. It is an electronic storage and retrieval device that provides spatial information, that is, it is an electronic atlas.

A GIS allows a user to see the patterns of settlement, to see landuse and natural resources of a water river basin, to search for their location and to identify relationships between the data. Specific features go beyond initial data types to include other features such as land systems (unique areas displaying recurrent patterns of terrestrial features), settlement features, best management options, land ownership and planning zones, demographic information and a range of other social and economic data.

GIS types vary. For river basin managers, a GIS should be a user-friendly system on the Web, which allows ease of access for all users for a specified and agreed number of purposes.

One of the critical factors in GIS use is to determine who owns and who will manage the GIS. This is a critical 'first step'. One option is to set up a corporately owned GIS between all stakeholders in a water management situation. A flexible design will allow users to adapt to new information needs as time progresses.

GIS can be used with visualisation technology. This allows the user to create 3D images of a water river basin and 'fly over' these images and view the landscape from different angles, in order to demonstrate changes in river basin landuse or environmental condition through time.

Another use of GIS is to georeference research and government reports. A further use of GIS is to track progress in the implementation of an IWRM strategy. Data in the GIS will demonstrate through time trends in the changes in resource condition.

Several key questions should be considered when using GIS:

- Needs' specification: What is the purpose of using GIS in the river basin? Has this been defined?
- Scale: Has the scale of application been determined and agreed on to suit all users?
- Agreements: Have opportunities been identified to create a corporately owned GIS? Have financial resources been targeted to find a corporately owned GIS?

- Data layers: Have the data layers been specified which suit the range of users of the GIS? Has there been an assessment of missing layers? What extra data sets are needed?

Specific cases of GIS include:

- The catchment resource assessment model is an information system that was developed in an attempt to provide decision-makers with data on competing interests within the Crocodile River Basin in South Africa. One of the state goals of this project was to develop a tool that would provide visual representation of spatial data in a format that could be utilised by the 'non-specialist user'.
- The Grand River Conservation Authority (Ontario, Canada) GIS has developed an in-house repository of river basin information, which will be used to assist their business operations, such as the need to improve levels of service (water supply and management), the need to reduce costs and the need for better coordination and management of the Authority's functions through stronger interdependence on GIS data (www.grandriver.ca).
- The Herbert River Information System uses a corporate GIS facility which involves collaboration between industry, the community, government and research agencies in the Herbert River (North Queensland, Australia) to specify cane land assignment for regional land and water management planning.
- The long-term hydrological impact assessment is a web-based decision support system developed at Purdue University (Indiana, USA). The framework of this DSS is to provide users with information pertaining to effects on water due to land-use changes. This project incorporates a GIS system.

Modelling and DSS

The terms modelling and DSS are becoming commonplace when discussing IRBM issues. The term model refers to a piece of computer software that mathematically represents processes we can see or measure on a field or across an entire river basin. However, these tools are often poorly understood. In many cases, there is a sense of mysticism surrounding models and DSS.

The term *Decision Support System* or *DSS* can be used in a variety of ways. A DSS is a means of collating and translating information or knowledge from a variety of sources to aid a water management decision. Sources of information can include experimental data, output from models, expert or local knowledge and information gathered from surveys. Most DSS can mix and match information from a variety of sources. A multi-objective DSS (MODSS) is a type of DSS that allows a user to integrate across a range of models and other data sets.

For the application of MODSS, five phases have been identified. Each phase can only be undertaken in consultation with all potential stakeholders.

- *Issue identification* where stakeholders, available information and possible consequences are identified;

- *Defining management options* where potential land management options are formulated;
- *Establishment of decision criteria* where the criteria that will be used to evaluate the different land management options are defined;
- *Data acquisition* where relevant information and data are collated and entered into the MODSS;
- *Decision support process* where the compiled information can be interrogated by a range of stakeholders with different points of view.

Models and DSS are best delivered if appropriate protocols are put in place first. These include distributing detailed model outputs in a World Wide Web compatible format (HTML) that can be accessed either across the Web, or on individual computers using any web browsing software. These model outputs can then be interrogated in a similar way that a user 'surfs the web'. A user can now search and interrogate model output in a more detailed form than ever before. Additionally, the actual modelling tool can/should be provided to a water management group including the library of model parameters and data. This must be in a user-friendly form.

This approach is transparent: nothing is hidden on computers back in the computer laboratory; all model inputs and model outputs are available for 'peer review'; and improvements are facilitated in modelling in future projects. The approach should ensure that future work can build on past research, that it does not start again from step one, that it prevents the value of the modelling activities being archived away on a backup tape; that it is likely to remove some of the mysticism of modelling; and that it takes river basin management groups' understanding of models beyond the 'black box'.

Several key questions should be considered when using DSS:

- Is one or both of these tools required to address some river basin management issue? More often than not, they will be because models can be used to fill gaps in existing information, and DSS can interpret information from different sources and facilitate decision-making especially for cases where conflicting objective cloud the decision-making process.
- Is access available to relevant expertise to provide guidance in the application of these tools?
- Have the appropriate delivery methods been worked out? Delivery mechanisms must be clearly established and defined prior to modelling activities, preferably at the project proposal stage.
- Is there 'after sales service' available after the project concludes?

An example is WAMADSS, a knowledge-based computer system that integrates data, information, physical simulation and economic analysis to identify alternative landuse maps for solving specific river basin problems. It identifies the relative contribution of sub-river basin areas to agricultural non-point source pollution and evaluates the effects of alternative landuse/management activities and practices on farm income, soil erosion and surface water quality at the river basin scale (Prato, Fulcher and Feng 1995).

There have been substantial efforts over the last decade to develop biophysical models that assist in the technical aspects of river basin management decision-making. These models have been developed primarily by and for resource management agencies at both local and regional scales. While it is impossible to draw up a comprehensive list of biophysical models given the large number of prototypes that have been developed to address specific situations, Table 5.1 identifies and describes a number of these models.

Classifying models

Reviews of biophysical modelling processes at the river basin level by (White, Yapp and Berry 1992; Shaw et al. 1996) indicate the increasing capability of biophysical modelling at the river basin scale. For example, Shaw et al. (1996) draws upon work by (Addiscott and Wagenett 1995) to identify three main approaches to modelling solute leaching: using deterministic models, stochastic models and physically based models. Deterministic models, where each type of process is considered to give uniquely defined outcomes, include mechanistic models and functional modals. The former are based on rate processes, incorporate the most fundamental of known processes, have generally been limited due to data requirements, are often based on laboratory experiments, and are useful to explore the implications of various model assumptions. Functional models, however, are based on capacity processes, give a simple summary of fundamental processes, with modest inputs, and have been developed for management purposes since data requirements are more modest. Stochastic models are applicable where outcomes are uncertain and the prediction of statistical limits of system response is required and has been developed to cope with highly variable environments. Physically based empirical models are used because of the difficulties of data and applications of deterministic models. They comprise concepts of more fundamental processes, but use simplified correlations and measured variables at more appropriate scales, both temporally and spatially, and are currently out of favour due to the narrow range of applications and extrapolation problems.

While these types of models have contributed to improving the technical and scientific rigour management, the above classification demonstrates the greatly focused nature of biophysical modelling. The sophisticated level of technical expertise required in running the models and in interpreting outputs has precluded community participation, ownership and understanding of the models. As a corollary, these models have limited potential use by community-based RBOs not trained in modelling technologies and demonstrate that appropriate protocols are needed to ensure effective adoption of model outputs by end users.

There have been few attempts to integrate social, behavioural and economic data (that explain impacts and adoption) into GIS-based DSS. Table 5.2 summarises

Table 5.1 Examples of biophysical DSS developed for natural resources management at the river basin and sib-basin levels

Name of model	Reference	Objective of application	Location of application	Description
Vegetation cover/erosion management DSS	Meijerink et al. (1993)	Developed scenarios for increased coffee production with regard crop suitability modelling	Komering River Basin, Indonesia	Combined remotely sensed data, relational database structure and field survey data to build a tool for solving questions about the management of coffee production
Aquatool	Andreu, Capilla and Sanchis (1996)	To develop tool for the planning and operational stages of dam operation	Spain	Comprises three control units: (1) optimisation and simulation models for basin management and aquifer flow; (2) two modules of risk assessment; (3) six modules for analysis of results
Waterware	Jamieson and Fredra (1996)	To develop comprehensive DSS for river basin planning	Thames Basin, UK Rio Lerma, Mexico	Integrates GIS, database management, modelling, optimisation procedures and expert systems in a modular format (i.e. components can be selected and incorporated)

IREM Integrated river basin environmental management system (IREM)	Integrated Decision Support Group (IDS) Centre for Water Resource Engineering and Science, Colorado State University	To provide for the evaluation, planning and management of wildlife habitat	Lonetree Wildlife Management Area, North Dakota	GIS to manipulate, evaluate and analyse spatial information; wildlife habitat models for indicator species to quantify quality of habitat; optimisation models for river basin management
RiverWare	Cadsweb.colostate .edu/riversware/ riverware_info.html		Colorado River, USA	

Table 5.2 Examples of integrated modelling procedures developed for resource management at the catchment level

Name of model	Reference	Objective of application	Location of application	Description
Open Modelling Engine	Young, Cuddy and Davis (personal communication)	To integrate models of different physical processes and different domains (e.g. socio-economic, biophysical models)	Australia	Modular construction which allows user to build up a representation of the system to be modelled
Influence	Walker and Young (personal communication)	To provide assistance in evaluating alternative landuse strategies	Australia	Describes systematically the relationship between various factors influencing catchment management to understand how the system will change in time and space
Several linear programming models	Oram and Dumsday (1996)	Integrate hydrogeological and economic information to model impacts of landuse change	Murray–Darling Basin, Australia	Used to evaluate social benefits and costs of salinity control
WAMADSS	Prato, Fulcher and Feng (1995)	To combine and evaluate in a spatial framework socio-economic and physical processes	USA	GIS used to display net social benefits and disbenefits of alternative management scenarios

SIRO-PLAN	Ive and Cocks (1983)	Provides an organised framework for acquiring and using information to analyse landuse options and balance demands of competing landuse interests in accordance with the judgements of clients	Thirty applications in rural shire/Local Government planning, New South Wales, Australia	Model comprises four steps: Establish terms of reference Collection of data Evaluation of plans Legitimisation, implementation and updating
LUPLAN	Ive and Cocks (1983)	Computer assisted evaluation of alternatives described in SIRO-PLAN: indicates changes in other policies from altering importance weights	Australia	Linear programming model which generates initial reference plan judged by client to be feasible then goes through a series of steps to improve the plan
Cross impact analysis (CAP) (forerunner of IDA)	Bonnicksen and Becker (1983)	Define a modelling system based on critically important variables and relationships between those variables	USA	CAP used panels of participants to estimate ratings and rankings for each of the variables; a cross impact matrix was developed which allowed participants to determine which relationships were used to produce normalised weighted sums for each variable

(*Continued*)

Table 5.2 Examples of integrated modelling procedures developed for resource management at the catchment level (*Continued*)

Name of model	Reference	Objective of application	Location of application	Description
Initial decision analysis (IDA) (forerunner of EZ-IMPACT)	Bonnicksen (1985)	Evaluate policies and programs	USA	Two satisfying techniques were used (i.e. mathematical and political satisfying) for evaluating policies
EZ-IMPACT	Bonnicksen (1985)	Judgment based program for incorporating public input into planning process	USA	Model comprises three modules: set priorities; construct models and perform experiments
Landscape change model	Lee et al. (1992)	Model which determines how socio-economic and ecological processes are integrated at the landscape level	USA	Conceptual model which develops an understanding of: landuse change that alters landscape patterns; effects of landscape patterns on species persistence, invasion of exotic species and resource supplies; dynamic interactions involving feedback processes that can alter uses or landscape patterns

Landuse change and analysis system (LUCAS)	Berry et al. (1994)	Incorporates socio-economic factors into a spatially explicit landscape change model	Little Tennessee River Basin, Olympic Peninsula, USA	Model generates maps which combine remotely sensed images, census data, land tenure, topographical information and outputs from econometric models. Simulations are generated which assess issues such as biodiversity conservation and landscape integrity
Integrated model of landuse and cover change	Reibsame et al. (1994)	Integrated model comprising four drivers of landuse change: human environment physical environment landuse decision processes ecological processes	Great Plains, USA	Conceptual model comprising four levels: physical environmental information agro-ecological modelling agricultural landuse models human environmental dimensions

(*Continued*)

Table 5.2 Examples of integrated modelling procedures developed for resource management at the catchment level (*Continued*)

Name of model	Reference	Objective of application	Location of application	Description
DSS for evaluating catchment policies	Davis, Nanninga and Clark (1989)	DSS which relates proposed landuse and management policies to effects upon selected water quality determinants	Mount Lofty Catchment, Australia	Model consists of two parts: model of catchment system and query component. Model of catchment comprises seven classes of object (policies, objectives, locations, land uses, determinants, time periods and interest groups). Messages passed between objects are determined by models that represent the best available information on catchment processes

P-DSS (multiple objective DSS for semi-arid rangelands)	Lawrence et al. (1997)	Uses measured data and expert opinion to quantify eight decision criteria in the evaluation of four management systems for semi-arid rangelands	Santa Rita Experimental Range, USA	P-DSS comprised of a decision model, a simulation model, an input file generator for the simulation model and a report generator. The P-DSS is used to select the preferred management system from four feasible grazing and vegetation manipulation systems. The evaluation incorporated eight decision criteria quantified using information from measured data sources and a survey of experts and five important orders
Long-term hydrological impact assessment	Engle and et al. (2003)	Integration of web-based programs and GIS	Indiana, USA	Consists of a modelling system, a graphical system and a database
Catchment resource assessment model	Chapman et al. (1995)	Simulation model based on changes in landuse	Crocodile River Catchment, South Africa	Runs model of land-use patterns, utilising water as 'currency'

some integrated modelling procedures which evolved in the 1990s and which are still useful to IRBM. These are discussed below.

New technologies are emerging to integrate economic and biophysical modelling. Young, Cuddy and Davis (CSIRO, Division of Water Resources, Canberra) (personal communication, 2000) developed software tools for use in a consultative river basin management process. Their 'Open Modelling Engine' is designed to integrate models of different physical processes and of different domains (for example, biophysical and socio-economic). They have developed skills in deciding what level of modelling complexity is appropriate to different user groups, in developing qualitative models, and in working with river basin management groups to determine their requirements and expectations with regard to software tools. Their plans for the 'Open Modelling Engine' are generally to:

- establish a system that allows the user to quickly build up a representation of the system to be modelled using a number of linked objects;
- allow the user to associate different models (either from a model library or user-defined) with these objects;
- provide the data handling routines (input–output and model-to-model);
- provide tools for rapid customisation of user interfaces for applications.

Davis, Nanninga and Clark (1989) developed a DSS that evaluates the effects of proposed landuse and management policies upon selected water quality determinants in the Mount Lofty Ranges of South Australia. This DSS is developed for evaluation of policies and is designed to respond to the diverse, exploratory questions posed by managers about a diverse range of land uses (e.g. urban, uncleared, horticulture, orchards, pasture and rough grazing). In order to answer queries, this 'Policy Analysis DSS' draws upon procedural (i.e. scientific) models, non-procedural (i.e. heuristic) knowledge and databases.

Walker and Young (CSIRO, Division of Wildlife Ecology, Canberra) (personal communication, 2000) developed the INFLUENCE software package, which systematically describes relationships between different factors influencing river basin management and provides a framework for systems-based modelling. In developing a land-use strategy for a region, it is important to understand how the region will change through time and across space under alternative development scenarios. INFLUENCE allows the tracking of decisions so as to determine net impacts. This procedure allows for the incorporation of the views and perceptions of stakeholders and presents these in the form of System Influence Diagrams. Keeping track of what factors are determining the agricultural land-use patterns and practices, and what direct and indirect impacts that result from these, can get very complex. The diagrams are a simple yet effective means of presenting the complex web of interactions in an easily understood manner.

Other technologies focus on predicting the costs of landuse change according to different land management units for a whole river basin. Oram and Dumsday (1996) have shown how hydrogeological and economic information can be combined to model the impacts of landuse change across a river basin. They developed a linear programming model to evaluate the social benefits and costs of salinity control. The model they developed is designed to link farm level results with river basin level off-site effects. A key function is to permit rapid evaluation of optimal landuse patterns and the sensitivity of management plans to assumptions concerning the off-site and downstream effects of groundwater recharge. It allows a method for justifying the financial support required for alternative management plans.

Prato, Fulcher and Feng (1995) maintained that the spatial information on socio-economic and physical processes needed for comprehensive evaluation of alternative river basin management plans is not readily accessible to local decision-makers. Consequently, they developed an interactive, watershed management decision support system (WAMADSS) that adopts a landscape perspective which views interactive parts of a river basin rather than focusing on isolated components. Their decision support system incorporates the use of GIS to give this landscape perspective. They have been able to provide a spatial display of net social benefits and detriments of alternative plans of management across a river basin.

In another prototype multi-objective DSS (P-DSS) for river basin management (Lawrence et al. 1997) integrated measured data and expert opinion to quantify eight decision criteria in order to evaluate four management systems (i.e. year-long and rotation grazing with the retention or removal of vegetation) for semi-arid rangelands. By considering the effects of alternative management systems on a number of decision criteria, decision-makers are presented with a ranking of the alternatives compared to the existing management system based on an importance order of the decision criteria.

Several earlier integrated, predictive models were developed, relevant to IRBM. LUPLAN (Ive and Cocks 1983) determined landuse scenarios according to a planning committee's changed importance weightings for different policies. SIRO-PLAN (Ive and Cocks 1983) was a stepped method to allocate landuse planning, which sought to satisfy interest groups and the degree of conflict between higher-level policies and regional policies for each policy under analysis. Later models that integrate data are the landscape change model developed in the Man and the Biosphere Program at the University of Washington (Lee et al. 1992) and the landuse change and analysis system (LUCAS) of Berry et al. (1994). Similar approaches were developed by Reibsame et al. (1994), who developed a conceptual model of landuse change for the Great Plains, integrating agro-ecological and economic models with anthropological characteristics of Great Plains' communities.

The Riebsame approach provides IRBM with an exploratory framework to identify the transitional probabilities of landscape change that could result from decisions made in IRBM programs. This approach could be constructed collaboratively by resource management agencies and regional river basin management committees. The model is particularly suited to predicting future ecological and social conditions, and for deriving procedures to redress impacts that flow from IRBM planning decisions.

Lee et al. (1992) described a further model, an analytical technique in which ecological processes and decision behaviour can be integrated in such a way as to predict impacts of landuse decisions on landscapes:

> Projecting regional patterns of landuse and land cover requires integrating the transitional probabilities of landuse change with the existing land cover patterns. Given a map of existing land cover patterns, the landuse transition probabilities are distributed spatially depending on the economic, social, and physical characteristics of each land parcel. The linkage of the probabilities to the spatially explicit database, such as is stored in a geographic information system (GIS), allows changes in a landscape through time to be simulated. Alternative scenarios can be explored, and the relative importance of different controlling factors evaluated.
>
> (p. 504).

The analysis of drivers of landscape change is a useful approach to understand past and current land management practices and predict future landuse change in a river basin by isolating significant drivers and analysing their importance to landscape change and consequent downstream water quality and quantity impacts. Models based on this approach require specificity regarding land types, land uses, landowner knowledge and resource use behaviour. These data exist, but models that effectively interact with the data on a GIS platform have had only limited application. It is essential that the model be transparent to allow wider application than site-specific empirically derived applications.

A further dimension of the landscape change model is to incorporate temporally dependent behavioural variables such as past, current and intended adoption behaviour of land and water management practices by resource users. A closer examination of the application of psycho-sociological, demographic and economic variables to the model is needed. Outputs from this type of model can be used to drive policy and program planning and implementation. This is best done using a stakeholder participation process such as adaptive environmental assessment and management (AEAM; Holling 1978; Walters 1986) to construct the components of the model.

5.2.3 Concerns about river basin management information systems

How effective are river basin IMS? What can be learned from using them? The following case study is a report of a survey[1] of the benefits and disbenefits of using these systems and the need for caution in their use. As part of a research on the development of a river basin information system discussed in Chapter 6, a survey was undertaken by the author on the use of the software catchment management support system, developed by CSIRO in Australia. The software, CMSS,

> is a tool for catchment managers and planners to examine the impact of policies, especially changes in landuse and/or land management practices, on water quality in catchments. It does this by predicting average annual loads of pollutants, usually Total Phosphorus and Total

[1] Undertaken as part of two broader studies of the use of information systems in river basin management (Hooper 1998; Hooper 2000) and the development of a prototype river basin information system (Hooper 1999).

> Nitrogen. CMSS is complemented by two other products: the Nutrient Generation Rate Data Book and NEXSYS, an expert system for predicting nutrient generation rates from different land uses under a range of climates and topography.
>
> http://www.clw.csiro.au/products/cmss/ accessed December 2004

Surveys were sent to a targeted population—The 59 licensees of the software at that time. With a 53 percent response, 76 percent responded that they had had an opportunity to use the software, leaving 24 percent without any practical experience applying the programme. The majority of those yet to use the software, planned to use it sometimes in the future; however some were doubtful citing changes in work direction and lack of commercial opportunities. Over half (54 percent) of those who will be using the software will use it to determine or monitor nutrient loads in the river basin and to assess the impact of land management use on the land. Other reasons were more general, being associated with the development of overall river basin management plans and more land oriented concerns such as rural/residential landuse and broad scale identification of problem areas.

Most respondents saw the responsibility of implementation as a coordinated effort involving local catchment management committees, government agencies and all affected stakeholders/landholders. Some, though, believed that it was entirely up to the associated government group or the catchment management committee (CMC). One responded that it was the role of the government to promote the outcomes and the landholders to implement them.

Again, most users/future users of CMSS Software believed that a number of clients benefit from the outcomes. Included in these were associated CMCs, local councils, government agencies and Landcare groups. Only around 30 percent included landholders as beneficiaries. A couple believed, in their experience, that no one benefited from the software.

Those with only future plans to use the software were unable to comment on the apparent strengths inherent with its use. Those respondents who had previously used the software were able to pinpoint a variety of strengths. 52 percent claimed that CMSS Software's greatest asset was that it was easy to run, use and understand, allowing landholders and community members to follow it. A few recognised its educational value as its greatest strength, claiming that it teaches the user that river basin management is a system made up of many parts which are interrelated. Other respondents said that it allows quantification for diffuse sources of nutrients, prioritisation and efficiency in achieving results and its adaptability to many situations.

There was little uniformity in the perceived weaknesses in CMSS Software. A few respondents agreed that results generated were too general and simplistic, making calculations based on averages and is only satisfactory for indicative purposes. One stated that the results are not sufficiently specific and credible to convince individual landowners to change management practices. Other weaknesses were believed to be:

- no expertise in its use;
- difficulty with large data sets;

- some screens and instructions were ambiguous;
- model can only calculate annual loads does not allow for seasonality;
- lacks ability to use GIS capabilities.

About half the respondents recognised that the outcomes predicted by the CMSS Software were tied in with nutrient levels for sub-river basins and nutrient contributions from various land uses. Others saw the outcomes as a broad river basin plan highlighting priorities, 'hotspots' and areas sensitive to further development. A couple of respondents saw the ultimate outcome as improvement in community liaison with regional council, research agencies and landholders.

Several respondents were unable to say whether CMSS Software had lead to a definite change in management practice due to it being too early in the process to tell. Of those who were able to respond, one-third stated that use of the software had not lead to any change in land management practices. This is because nutrient management and other plans were still to be implemented, and that results were inconclusive and not definitive. Of those who found that the software has brought about changes in land management, these changes were achieved through nutrient management plans and water quality strategies based on CMSS results.

When asked what hinders the effectiveness of using the outcomes of CMSS, quite a number of respondents expressed doubt over the reliability and accuracy of the outputs, generally due to its 'broad brush approach' to river basin management. However, a couple of users maintained that the CMSS Software is only a tool to aid and assist in the initial development of management options, i.e. in decision-making, and that problems in its effectiveness occur due to users assuming that the model will decide management practices for them and calculate accurate costs and nutrient loads. Additional responses included:

- lack of confidence in results;
- needs an extension follow up for farmers;
- not portable;
- data hungry;
- does not provide information for questions being asked;
- too generic to influence individual land managers;
- land managers resistance to change;
- difficulty in getting CMSS expertise in New Zealand.

Respondents were asked to comment on possible hindrances to the implementation of the software outcomes by landowners or Local Government. The results were quite diverse, with little similarity in the given reasons. However, a number of respondents stated that 'farmers do not care' among their replies. Responsibility and ownership issues with regard to water quality problems were another acknowledged hindrance. Other responses included:

- no incentive to change;
- reluctance to accept they are part of the problem;

- general disbelief in the outcomes of a computer model;
- a need to be shown how CMSS information fits into the decision-making/legal framework;
- cost implications of improved management (economic considerations);
- communication problems between land managers and the agents undertaking CMSS studies;
- lack of legislation and other forms of coercion;
- failure to include from the outset all relevant people affected;
- communication problems, not model software.

The most significant finding of this study is that current catchment management planning processes appear to mitigate against the effective use of computer software to assist catchment management decision-making. Many of the concerns about end users not implementing software outputs are related to the lack of effective catchment management planning, not the software available to support it.

Another outcome of the survey was the repeated belief that many weaknesses in the implementation of the outcomes and in the general effectiveness of the software was due to poor understanding of the results by farmers, and an apparent lack of interaction between stakeholders and various committees and agencies. Reasons such as 'needing an extension follow up for farmers', 'the general disbelief in the outcomes of a computer model' illustrate that such decision-making software as CMSS is lacking a vital 'social' component that incorporates farmers' attitudes and behaviours, farmers' understanding of issues and the economic characteristics of farming. It appears that most hindrances were not found in the software itself, but rather between the various users.

There appears to be a lack of effective, practical implementation processes whereby software users can pick up the outputs and implement them. Therefore, even though it might aid catchment management committees in decision-making, unless farmers can understand and trust it, very little will change in terms of land management practices. It is therefore incumbent on users of software like CMSS to have a catchment management planning process that has clearly defined implementation processes designed to take the output of a DSS and use it. It is not the role of the software to do this, rather the role of the managers who use it.

5.2.4 The concept of an integrated information management system

From this discussion, it can be seen that there are limitations to the application of modelling technologies, GIS, database inventories and expert systems in river basin management. The prime concern is how such technologies relate to river basin organisations which are becoming increasingly driven by strong public involvement. Care is needed not only to choose the correct model, but to recognise the limitations of available data sets, to know the context of the decision situation and the needs of the modelling tool and the users, but also the role of resource managers 'on the ground' (farmers, foresters and the like) have in the use of information exchange for river basin management.

Another approach to information management by river basin organisations is needed: one that allows more ownership, operation and use by a broad range of stakeholders. One option is to use an 'Integrated Information Management System' (IIMS); a term used to describe a system to organise, coordinate and assist in the dissemination of knowledge about social, economic and biophysical processes in a river basin. This knowledge includes the output of research projects, government information and many types of community and individual river basin knowledge.

The role of an IIMS is to facilitate the dissemination of information useful to resource management decision-making at the sub-basin and river basin level and assist in the selection of best management options from the most current and realistic information available. An IIMS must address information flow pathologies. To be effective, it needs to develop a process whereby information and data of a scientific, technical, social and economic nature can be integrated and employed in such a manner that well-informed, timely, participatory and strategic decisions can be made in relation to natural resources management.

The development and testing of a prototype IIMS, called a river basin management information system (RBMIS), is discussed in Chapter 6.

5.3 STAKEHOLDER-DRIVEN RIVER BASIN MANAGEMENT INFORMATION EXCHANGE

While there is a wide range of information tools, there are relatively few that are designed specifically for the diverse array of stakeholders in river basin management. This is because various systems have evolved according to different needs.

Natural resources management is becoming increasingly complex as a result of rapidly changing political, social and economic conditions, and the growing range of data and information sources. In relation to inadequacies of the existing decision-making process, Walker and Johnson (1996) observed that river basin management is characterised by the occurrence of occasional or one-off tasks rather than routine, frequent and standardised processes. The use of complex modelling procedures that provide only partial decision support and lead to 'black box' decision-making and low levels of stakeholder and community participation can result in distrust of resource management recommendations and, as a corollary, delay the adoption of improved management practices.

There are also concerns that 'black-box modelling' is removed from the information flow paths between researchers and river basin management committees, government agencies and individual farmers. This issue is discussed later in this chapter. There is a need to link models that evaluate resource use options to the prevailing information flow paths in the social organisation of river basin management decision-making. This will have two benefits: increased ownership by clients and users of modelling outputs, and improved quality of modelling by using the end users in model design.

5.3.1 Building capacity for improved decision-making with information

The assessment of resource management options is required by a growing variety of clients including landuse planners, ecologists, environmentalists, economists, agricultural extension facilitators and land users themselves. In a watershed, the needs of these clients vary widely and include issues such as land suitability and productivity assessment, quantification of land resource constraints, agricultural technology transfer, agricultural input recommendations, farming systems analysis and development, agro-economic zoning for land development and nature conservation and ecosystem research and management (Sombroek and Antoine 1994). Accordingly, the core challenge in designing and constructing an information exchange system to address the fragmentation of the decision-making process and the variety of information outputs.

Within this context, there are calls for decision-making to be based on comprehensive assessments of development and management alternatives which take into account biophysical, environmental and socio-economic factors as well as the time and space dimensions of sustainable land management (Reibsame et al. 1994; Sombroek and Antoine 1994; Shaw et al. 1996).

Well-designed and constructed IIMS which combine scientific and technical knowledge with socio-economic, political and cultural information can offer considerable benefits in the management of land and water resources. These benefits include:

- the drawing together of a large number of disparate information sources of a scientific, engineering, socio-political and economic nature which can provide land owners, institutional and elected representatives with timely, integrated information which is essential for good planning and resource management;
- an increase in the systematic use of information which enhances the rationality of the decision-making process;
- the ability not only to predict but to provide expert advice on appropriate development and management alternatives thereby improving the current management process from a reactive to a strategic one;
- increased transparency of the decision-making process whereby the effects of various options can be identified and impacts can be openly assessed in a form readily understood by the non-expert;
- the facilitation of information transfer which results in education and increased awareness of all participants involved in resource management decision-making processes;
- the evaluation of untried management alternatives which is useful in circumstances where traditional practices are more likely to be adopted than new, innovative approaches because of lack of awareness or fear of the unknown.

5.3.2 Participatory approaches for the development of an IIMS

Research in the area of extension science suggests that new complex methods and management procedures are unlikely to be accepted and acted upon if the end-users are unable to understand or interpret the problem and its implications (Hollick 1990). Lee et al. (1992) have called for the formation of 'ecologically effective social organisations', that is, social structures and protocols to link end-users with resource management agencies and other government planning agencies (particularly Local Government) to achieve more effective decision-making. Johnson and Walker (1997) demonstrated by way of needs analysis and cost benefit analysis that the development of a corporate GIS facility that involves collaboration between industry, the community, government and research agencies in the Herbert River (North Queensland, Australia) is a prudent investment. It was argued that the development of this corporate GIS will result in greater ownership of resource management planning by both agencies and resource managers, which in turn would increase the potential for implementing ecologically sustainable development.

New participation methods and technologies have been developed that attempt to increase participation of stakeholders and end-users in river basin management. Among these are interacting group methods, workshops, social surveys, Delphi techniques (including AEAM and microcomputer-enhanced decision-making techniques such as SIRO-PLAN, Cross Impact Assessment and Initial Decision Analysis (Bonnicksen 1985) that attempt to use community and expert advice to improve decision-making (Stark and Seitz 1988). These models use both expert knowledge and local wisdom about resource use to be included in the construction of models to explain river basin processes.

Grayson, Argent and Ewing (Centre for Environmental Applied Hydrology, The University of Melbourne) (personal communication) developed flexible methods and a number of tools to improve the integration of land and water R&D at the river basin to farm scale of sub-basin management. They used an adaptive environmental assessment method and associated workshop and modelling procedures. The purpose and design of their project was focused not on the modelling task itself, but rather on the effectiveness of that process in furthering the integration of research and development technologies and the subsequent adoption of improved management practices.

There are a range of mechanisms and tools available to allow information exchange, according to different settings:

- One-to-one exchange using telephone and fax services and exchanges during social experiences, conferences, symposia and professional meetings;
- Text material: newsletters (paper and electronic), printed manuals, newspaper and electronic media reports, bulletin boards, email chat and email correspondence about IWRM experiences;
- Interactive web-base watershed information systems that specify best management options for land types (land systems, land management units, planning

areas, sub-watersheds) in a river basin and that are congruent with overall river basin management goals and targets;
- Interactive GIS built for use within agencies or targeted partners in a water management context;
- Field days, farm demonstrations, Landcare[2] group meetings and workshops to exchange best management practice experiences at the local level;
- Professional workshops to exchange experiences in state-of-the-art tools;
- Radio broadcasts and video presentations in regions of low literacy;
- Village level capacity building through one to one discussion with farmers and village leaders;
- National and regional technical tours allowing professionals and practitioners to exchange first-hand results of IWRM, using discussion in the field.

The lessons learned from using these information exchange methods include:

- Word of mouth networks form powerful mechanisms for information exchange in IWRM.
- The Internet should be used cautiously due to limited access and expense in some countries and rural settings.
- Electronic exchange mechanisms are inappropriate in most settings unless usage protocols have been agreed on by partners in a water management context.
- Traditional top–down extension is limited: users and providers must engage one another and build partnerships and trust for effective information exchange.
- Like watershed management, champions drive effective information exchange.

Specific cases include:

- The Great Lakes Information Network (GLIN) described as 'a partnership that provides one place online for people to find information about the bi-national Great Lakes region of North America'. GLIN has synthesised a large volume of contact information into simple and visually appealing Websites that hotlink readers easily to other locations (http://www.great-lakes.net/).
- The Chesapeake Bay Information Network has developed a gateway to a vast array of Internet resources, providing sub-river basin descriptions, access information to Federal and State Government programs, events listings, organisation contacts, funding opportunities and other environmental networks (www.chesapeake.org).
- Mayfield (University of Waterloo, Canada) developed an open access information system for subwatershed development and planning. It is a community information system whereby scientists and all players put information into a common database. All players buy into the concept though cash or in-kind contributions. It incorporates in a presentation mode all ecological

[2] Landcare is a process whereby farmer groups undertake local catchment management actions.

zoning, eco-zones, bioregions and an overlay of significant species. Landuse planning consultants can use the system as it presents constraints and opportunities for land uses. References to planning acts, laws of Ontario, river basin plans and updates to the data can be sent to developers and consultants (http://bordeaux.uwaterloo.ca, accessed November, 2004).

5.3.3 Principles of information exchange

Stakeholders in river basin management learn, and information is exchanged better, if they meet face to face, learn from each other, mentors or champions, and discuss its application one-to-one or in interest groups. If enhanced information exchange is to be achieved, these professionals need a process that must be:

- *Appropriate*—providing information that is relevant to the task at hand, has been tested in the field and rigorously proven through research and development, and applicable to the type of problem, the level of institutional capacity and technical ability of the practitioners, and where these characteristics are limited, mechanisms are available to easily make the exchange process work in the short term.
- *Affordable*—by resource managers, preferably free, so that information providers and users are not discriminated against due to lack of funding.
- *Accessible*—to all practitioners in methods they frequently use with the information exchange process not relying on major upgrades in individual and organisation technical ability but fit their current information exchange processes.
- *Equitable*—the information process should embrace an expanded mandate which is sensitive and respects cultural needs and gender issues and does not discriminate against users and providers because of their distant location from the decision-making processes of other water professionals.

Reorienting information services for IRBM

The responsibility for managing and exchanging information for river basin management is shared among individuals, community groups, water management professionals, natural resource management institutions and governments. They must work together towards defining and implementing information management protocols and tools which can contribute to sustainable river basin management.

5.3.4 Principles of a River Basin Management Information System (RBMIS)

The development of an RBMIS is best done using participatory techniques. In this chapter it has been shown that few systems have been developed for stakeholder-driven information systems. The following principles are suggested as the basis upon which such an information system should be constructed:

- The RBMIS should be based on *capacity building* (including the empowerment and skilling of community organisations) in addition to being technology driven.
- The RBMIS should be *demand-driven* whereby the needs of the resource managers are clearly articulated at the outset of system design and construction and the outputs are directed toward these end users.
- The RBMIS should be *flexible* enough so that the system can be transported and implemented in a variety of locations.
- The RBMIS should be *transparent* and *rigorous* in the way it processes information so than technical and non-technical persons can follow the process of alternative generation and evaluation.
- The RBMIS should be *interactive*, which facilitates a participatory decision-making process.
- The RBMIS should *facilitate the process of learning* and *increase awareness* of the various management and development options available.
- The RBMIS should have the capacity to *integrate multidisciplinary information*.
- The RBMIS should be *readily accessible* to the large number of stakeholders and participants in the resource decision-making process.

The mechanistic, model- and technology-driven approaches described earlier in this chapter present a number of problems when engaging rural and regional communities involved in river basin management. The problems include:

- Methods of handling of social and economic information, including how social and economic information should be integrated with biophysical information to provide decision-makers with clear understanding of the constraints and opportunities to implement land management options thought to lead to more sustainable landuse.
- Who defines and how so-called 'best' management practices are defined;
- Difficulties and reluctance to use knowledge of local attitudes, beliefs and values about resource use and how they influence program implementation;
- Limited use of local farmer knowledge, anecdotal information, indigenous land management knowledge and how these data sets can be used in decision-making with contemporary scientific knowledge;
- Lack of knowledge of information flows and decision-making in a river basin and how these processes affect river basin management and the use of information;
- The role of institutional arrangements (laws, policies, agreements, incentives, taxes) for river basin management that may impede information flows at the river basin level. These arrangements are critical to the success of river basin management as well as information management in a river basin.

A process is needed to allow easy access to simplified accurate information relevant to a range of decision-makers. Designing a process of information transfer and exchange is not easy and needs to involve all relevant players. There is the need to have a range of interpreted natural resources data and information about natural resources management, applicable to a range of river basin managers. These include critical use limits and targets to achieve sustainable natural resource management. The information is also needed at a scale appropriate to river basin scale decision-making.

In a workshop held to define these river basin management information problems, agreement was reached by a group of ten leading researchers and practitioners in Australian river basin management. The workshop was instigated by the Australian Government's Land and Water Resources Research and Development Corporation and was held in Brisbane in 1996. The workshop concluded that,

> river basin and other resource managers do not have the information and skills (data, knowledge, models, and human resources) to integrate and to adequately respond to natural resource management issues, especially with respect to appropriate resource use. In addition, the volume of information is increasing, the reliability of the information is not being tested, and there are few processes in place to deliver the information to end-users. Furthermore, there are uncertainties regarding the institutional arrangements governing river basin management organizations (RBOs) and their role in information dissemination. It is imperative that decision-making by RBOs is understood before effective information management systems can be designed to assist them.

The result of this workshop was the development of a prototype river basin information exchange programme and this is discussed in Chapter 6.

5.4 SUMMARY

Several new technologies to deliver river basin information have been discussed in this chapter. These include Web-based information serves, DSS and GIS. Technological innovation is strong, progressive and innovative in these three areas. Many state of the art technologies have been discussed, for example DSS for riverine management (www.cadsweb.colorado.edu/riverware/riverware_info.html). However, many of these technologies were developed for specific client driven requests, where the clients were often government agencies. Consequently the interpretation of outputs by community organisations and individuals is difficult as they may lack technical expertise, or be unable to access the technologies used by the agencies.

> Technological innovation is widespread, particularly GIS and DSS technology, but it is developed and used mainly by technicians, and less frequently by planners and other decision-makers.

The solution to this problem is to develop technologies that synthesise and integrate information according to the needs of river basin decision-makers. There is increasing public involvement in river basin management in many countries and it is logical to frame questions about information delivery on a user-needs basis.

In this chapter we have seen that information systems for river basin managers are relatively new, but rapidly expanding. Integrated modelling technologies for river basin management are useful to predict the social and biophysical impacts of alternate management plans in river basins. The challenge remains to develop information systems with full stakeholder involvement, that perform modelling, decision support and visualisation functions, that are owned and operated by river basin management organisations, and that are robust enough to accommodate changing environmental and social conditions. There is an imperative to build integrated information management systems that include appropriate protocols and that assist rather than detract from emerging community-driven decision-making processes.

5.5 REFERENCES

Addiscott, T. M. & Wagenett, R. J. 1995, Concepts of solute leaching in soils: A review of modelling approaches. Journal of Soil Science 36, 411–424.

Andreu, J., Capilla, E., & Sanchis, E. 1996, AQUATOOL: A generalized decision-support system for water resources planning and operational management. Journal of Hydrology 177, 269–291.

Auscher, R. 1993, Microcomputers in support of the generation and diffusion of agricultural technologies in developing countries. Outlooks on Agriculture 23[4], 147–150.

Bauer, K. W. 1996, Creating integrated rural resource land information systems. Journal of Soil and Water Conservation 1, 29–33.

Berry, M. W., Flamm, R. O., Hazen, B. C., & Macintyre, R. M. 1994, The Land Use Change and Analysis System (LUCAS). Seattle, WA, USA, University of Washington, Center for Streamside Studies.

Bonnicksen, T. M. 1985, A participatory approach for developing resource policies. Environmental Management 9[5], 379–392.

Bonnicksen, T. M. & Becker, R. H. 1983, Environmental impact studies: An interdisciplinary approach for assigning priorities. Environmental Management 7[1], 109–117.

Chapman, R. A., Manders, P. T., Scholes, R. J. and Bosch, J. M. 1995, Who should get the water? Decision support for water resource management. Water Science and Technology 32[5–6], 37–43.

Davis, J. R., Nanninga, P. M., & Clark, R. D. S. 1989, A Decision Support System for Evaluating Catchment Policies. Melbourne, Australia.

Engle, B. A., Choi, J-Y., Harbor, J. and Pandey, S. 2003, Web-based DSS for hydrologic impact evaluation of small watershed land use changes. Computers and Electronics in Agriculture 39[3], 241–249.

Fredra, K. 1996, Decision support systems for natural resources management: Models, GIS and expert systems. http://iiasa.ac.at//Research/ACA/papers/toronto.html.

Hamilton, W. T. & Sheehy, D. P. 1993, Introducing a decision support system to agriculturally developing countries: An example from North China. Stuth, J. W. & Lyons, B. G. (eds.). Decision Support Systems for the Management of Grazing Lands: Emerging Issues. 221–253. New York, NY, USA, UNESCO/Parthenon Publishing Group.

Hollick, M. 1990, Land conservation policies and farmer decision-making. Australian Journal of Soil and Water Conservation 3[1], 6–13.

Holling, C. S. 1978, Adaptive Environmental Assessment and Management. Vol. 3. Chicester, Wiley, Wiley International Series on Applied Systems Analysis.

Hooper, B. P. 1998, Delivering R&D to Catchment Managers: A Review of Selected North American Experiences and Implications for Australian Practice. UQL15. Brisbane,

University of Queensland. Research and Development Final Report (Land and Water Resources Research and Development Corporation).
Hooper, B. P. 1999, An Integrated Information Management System for Catchment Managers. UQL10. Integrated Resource Management Research Pty Ltd. Research and Development Final Report (Land and Water Resources Research and Development Corporation).
Hooper, B. P. 2000, Catchment Information System: Distribution and Publication. IRM2. Brisbane, Integrated Resource Management Research Pty Ltd. Research and Development Final Report (Land and Water Resources Research and Development Corporation).
Ive, J. R. & Cocks, K. D. 1983, SIRO-PLAN and LUPLAN: An Australian approach to land use planning 2. Environment and Planning B 10, 347–355.
Jamieson, D. G. & Fredra, K. 1996, The "WaterWare" decision-support system for river basin planning 1: Conceptual design. Journal of Hydrology 177, 163–175.
Jamieson, P. 1994, Johnstone River Catchment. Queensland, Australia. Integrated Catchment Management Conceptual Interactions Model. University of New England. M.Sc. Thesis.
Johnson, A. K. L. & Walker, D. H. 1997. Development and Implementation for a Corporate GIS in a Coastal Rural Catchment in Northeastern Australia. Unpublished Work.
Laut, P. & Taplin, B. J. 1989, Catchment Management in Australia in the 1980s. CSIRO Divisional Report 89/3. Canberra, CSIRO.
Lawrence, P. A., Stone, J. J., Heilman, P., & Lane, L. J. 1997, Using Measured Data and Expert Opinion in a Multiple Objective Decision Support System for Semiarid Rangelands. Transactions of the ASEA, 40[6], 1589–1597.
Lee, R. G., Flamm, R. O., Turner, M. G., Bledsoe, C., Chandler, P., Deferrari, C., Gottfried, R., Naiman, R. J., Schumaker, N., & Wear, D. 1992, Integrating sustainable development and environmental vitality: A landscape ecology approach. Naiman, R. J. (ed.). Watershed Management: Balancing Sustainability and Environmental Change. 499–521. New York, Springer-Verlag.
Meijerink, A. M. J., Mannaerts, C. M., De Brouwer, H., & Valenzuela, C. R. 1993, Application of ILWIS to decision support in watershed management; case study of the Komering River basin, Indonesia. Kovar, K. & Nachtnebel, H. P. (eds.). Application of Geographic Information Systems in Hydrology and Water Resources Management. Louvain, Belgium, International Association of Hydrological Sciences.
Oram, D. & Dumsday, R. 1996, Socio-economic Analyses of Dryland Salinity Control in the Murray–Darling Basin. Project Status Report. Gunnedah, Liverpool Plains Land Management Committee.
Prato, T., Fulcher, C., & Feng, X. 1995, Decision Support System for Total Watershed Management. Research Report 14. Columbia, Missouri, USA, Center for Agricultural, Resource and Environmental Systems, University of Missouri-Columbia, Missouri.
Reibsame, W. E., Parton, W. J., Galvin, K. A., Burke, I. C., Bouhren, I., Young, R., & Knop, E. 1994, Integrated modelling of land use and cover change. Bioscience 44[5], 350–356.
Reitsma, R. F. 1996, Structure and support of water-resources management and decision-making. Journal of Hydrology 117, 253–269.
Shaw, R. 2003, The Role of Models in Decision-making for Natural Resource Use and Management. Australia, Murray–Darling Commission Workshop.
Shaw, R., Doherty, J., Brebber, L., Cogle, L., & Lait, R. 1996, The Use of Multi-Objective Decision-Making for Resolution of Resource Use and Environmental Management Conflicts at a Catchment Scale.
Sombroek, W. G. & Antoine, J. 1994, The use of geographic information systems (GIS) in land resources appraisal. Outlooks in Agriculture 24[3], 249–255.
Stark, C. R. & Seitz, W. D. 1988, EZ-IMPACT: A judgment-based program for public input into resource management and modelling. Society and Natural Resources 1[4], 387–395.

Tecle, A. 1992, Selecting a multicriterion decision making technique for watershed resources management. Water Resources Bulletin 28[1], 129–140.

Tolley, G. S. & Riggs, F. E. 1961, Economics of Watershed Planning. Ames, Iowa, USA, The Iowa State University Press.

Walker, D. H. & Johnson, A. K. L. 1996, Delivering flexible decision support for environmental management: A case study in integrated catchment management. Australian Journal of Environmental Management 3[3], 174–188.

Walters, C. J. 1986, Adaptive Management of Renewable Resources. New York, NY, USA, Macmillan Publishing Co.

White, D., Yapp, G., & Berry, G. 1992, Regional and catchment modelling. Agricultural Systems and Information Technology 42[2], 1–45.

6

A prototype river basin information exchange programme

6.1 DELIVERING R&D TO RIVER BASIN MANAGERS

6.1.1 Understanding research adoption

An issue of increasing importance in river basin management is the uptake of research outcomes, specifically research on new resource management methods. The question is how can this information be used to assist decision-making? There is often weak adoption of research outcomes in river basin management programmes. River basin organisations are not equipped to handle (integrate, synthesise, interpret and disseminate) research outcomes.

There are four critical dimensions of this research adoption problem:

- There is an increasing amount of information becoming available to river basin and other resource managers, but they lack the skills and techniques to adequately respond to natural resource management issues and convey appropriate landuse options to end users.
- The volume of information is increasing, the reliability of the information is not being tested and there are few processes in place to deliver the information to end-users.
- There are uncertainties regarding the institutional arrangements governing river basin management committees and their role in information dissemination.

- It is imperative that decision-making and information flows in a river basin are understood before effective information management systems can be designed to assist river basin management organisations.

These dimensions suggest that there are problems in integrating data and integrating research results. There are institutional, organisational and technical factors precluding the adoption of research.

Hooper (1998) reviewed Canadian and US experiences in developing and implementing stakeholder-driven information management systems for river basin and local watershed management. The study examined systems that integrate social, behavioural and economic data into Geographical Information System (GIS)-based decision support systems and assessed the relevance of these experiences to river basin information exchange. The focus of the study was to work out the best ways to deliver natural resources management research to river basin managers. A sample population of 63 river basin managers, researchers and government employees were used to analyse research and development delivery. They were members of river basin management organisations, R&D funding organisations, community-based/citizen river basin advocacy groups, Local Government organisations, State and Federal Government agencies and research institutions. The last group focused on R&D extension programmes in American universities.

Several successes and failures were identified in the study. Much has been achieved using extension programmes to increase on-farm adoption of best management practices. Very large extension programmes have been developed in universities, but these are still coming to grips with the watershed approach. Maturing Canadian river basin organisations and the conservation authorities are struggling with information delivery issues, yet they have achieved governance and respect as regional organisations able to deliver river basin management. Considerable technological innovation has occurred in building powerful decision support systems and geographic information systems. However, three fundamental problems were found:

- We still know little about how best to deliver R&D to river basin management decision-makers.
- There is little integration of R&D output that provides river basin management decision-makers with resource use options.
- There has been widespread lack of involvement of end-users in information delivery processes, yet they are the people who will use the R&D.

The remainder of this chapter discusses the results of the study and provides guidelines and a prototype for improved exchange between researchers, river basin organisations and resource managers.

6.1.2 The need for integrated information systems

The overall finding of this study was that R&D delivery to river basin managers was a universal problem in Canada and the USA (Hooper 1998). Although the scales of river basin and catchment management varied, the problem experienced by many practitioners was the failure to implement research because institutional barriers mitigate against effective R&D delivery at the river basin scale. Where it was successful, it occurred because an institutional arrangement had been created to deliver research products directly to end-users, and these recipients helped frame research questions. An excellent example is the work of extension programmes in American universities, such as the University of Wisconsin-Madison extension programme in water quality. The programme delivers best practice information to farmers on a one-to-one basis, using water quality education programmes based on information manuals. Farmers, in turn, assist framing the research questions. As a result, water quality improvements have been made. This is not a catchment management programme, rather individual farmer adoption programme. The UWEX programs have recently adopted a watershed, for example, the Rock River 2000 Demonstration Project (http://www.uwex.edu/ces/ag/sus/rock.html). The UWEX program is diverse with strong emphasis on delivering community and ecological sustainability, building on the success of decades of individual farmer adoption extension.

> Delivering R&D to assist river basin management is a widespread problem and few integrated information systems have been developed.

6.1.3 The nature of river basin management research

There are significant problems in understanding how river basin management research products should be delivered to river basin managers. Many interviewees commented on the decline of traditional research and university extension programmes, and maintained that a new R&D paradigm is emerging. This new paradigm incorporates end-users in research design. The imperative is to harness end-user involvement from the outset of a research project and to guide the management of the project within a river basin management-planning context. Narrowing the gaps between the priorities of research, funding organisations, researchers and end-users will ensure improved research uptake.

> The focus of IRBM needs to be reframed—focussing on end-users needs and stakeholders engagements.

Some interviewees maintained that this approach challenges the way researchers do their work, as there is no incentive structure in place in universities and other research institutions to work with end-users. It is more important to publish research results in scholarly journals to gain professional advancement. The latter is the incentive.

Some researchers, particularly those in the prime of their research careers enjoy the new paradigm. As one interviewee said, ‘they have little academic credibility to lose’. Younger researchers may see it differently and fear they will reduce opportunities for further academic advancement if they cannot keep up a substantial rate of high quality publications.

The new research paradigm is difficult to define because, like IRBM, it is emerging and, in many respects, is a social experiment. One interviewee maintained that it should be seen as a stepped process, or series of research questions requiring fundamental answers, with each being a step in the research programme of a river basin management-planning process (S. Born, University of Wisconsin-Madison, personal communication, November 4, 1997):

- What is happening to river basin ecosystems?
- What are the forces that cause things to happen in and to ecosystems by resource use practices?
- What are the interactive effects of these forces?
- Do we have enough confidence in our models to know that if specific variables are changed, these are the outcomes?
- Are these outcomes desired and are they part of the social decision system of the river basin?

These steps include studies of the contents, processes and interconnections within and between ecosystems and modelling the predictive capability of ecological change.

> Knowing this information, we then seek solutions to resource management problems. These are based on the predictive capability of science. We run tests to understand causality and do this by testing various management options, and for each, determine the degree of confidence one can ascribe to each solution. In undertaking these tasks one needs to target data needs, and we find frequently that we need everything. The problem then is to reduce knowledge, data, and information down to a size that provides explicit options for resource use to resource managers.
> (Born, personal communication, 1997)

Knowledge of the social system is the key requirement to improving the likelihood of the adoption of research results. This involves undertaking a social and demographic survey in a river basin to identify the resources of the river basin. S. Rikoon (personal communication November 27, 1997) refers to these resources as:

- Human capital: capability (skills, abilities) and leadership within the river basin community.
- Social capital: networks of information flows, multiple conduits of information flows, information providers and users, knowledge systems, informal systems of information flow, associations and other civic institutions.
- Cultural capital: existing attitudes, beliefs and values that guide behaviour.

The approach to research in a river basin management context suggests that the functionality of an information delivery system for river basin managers should include:

- information and data on the river basin,
- access to assistance that helps understand problems,

- descriptions of interventions[1] to achieve sustainable resource use, designed with cooperative action with end-users (and geo-referenced to land systems in a river basin)
- reports on the successes and failures of field testing of interventions.

6.1.4 Dimensions of information delivery

There are many techniques which were being used to deliver information about natural resources, natural resource conditions, management practices and research outputs in a river basin. These include (not in rank order of importance):

- *One-to-one farmer traditional extension programmes*, including the use of best management practices micro-targeted to the physical conditions of and financial conditions over individual farm operations. They are more likely to be effective when the socio-economic and biophysical conditions of the individual farm operation are met (for example, the UWEX approach discussed earlier).
- *Web-based interactive information management systems* that geo-reference information. These include GIS-based river basin information at the regional scale or more localised scales. Such systems have been developed at CARES, University of Missouri (http://www.cares.missouri.edu/).
- *Widespread publicity campaigns* using newsprint, television, radio and Web-based media. These approaches provide specific information, but it is difficult to target the research output to individual decision-makers because of the need for general relevance. This approach has been favoured by many government agencies, conservation authorities and river basin groups to harness public interest in IRBM.
- *Specialised training programmes* for river basin managers. The outstanding example of this approach is the *Know Your Watershed* program, run by the Conservation Tillage Information Center, linked to Purdue University, Indiana. This organisation runs a comprehensive programme in training watershed management skills with an emphasis on facilitation and empowerment skill of local communities (http://www.ctic.purdue.edu)).

River basin management information delivery is a composite process and should therefore use several mechanisms simultaneously to deliver research products to decision makers.

These observations suggest effective R&D delivery is a composite process.

[1] These are scaled down interventions that will give a better bounding of the natural resource management problem in the catchment. An adaptive management process is used to field test the interventions in a small area. Then methods are worked out to extend the outcomes to the catchment scale, and in so doing, work out what variables are most significant at this larger scale—variables in the biophysical, and socio-economic systems in the catchment.

6.1.5 River basins as information 'clearing houses'

There is little integration of different forms of information delivery. Most efforts are based on publicity campaigns and placing materials on the Web with little interactive capability. However, a common approach to providing information to both rural and urban communities is to use the Web as a river basin 'clearing house'. The Grand River Conservation Authority, for example, developed an in-house repository of watershed information which will be used to assist their business operations, such as the need to improve levels of service (water supply and management), the need to reduce costs, and better coordination and management of the Authority's functions through stronger interdependence on GIS data (www.grandriver.on.ca).

> Current practice in information delivery to communities emphasises the Web, publicity campaigns and traditional farmer extension—information needs to be relevant to the context and link to an overall river basin management approach.

On a much larger scale, the Great Lakes Information Network (GLIN) described as 'a partnership that provides *one place* online for people to find information about the bi-national Great Lakes region of North America'. GLIN has synthesised a large volume of contact information into simple and visually appealing Websites that hotlink readers easily to other locations (www.great-lakes.net/). The (bilingual) programmes homepage in GLIN describes key regional, national and international programmes that are readily available for community access.

On a similar large scale, the Chesapeake Bay Information Network developed a gateway to a vast array of Internet resources, providing sub-river basin descriptions, access information to Federal and State Government programmes, events listings, organisation contacts, funding opportunities and other environmental networks (www.chesapeake.org).

The development of the North Carolina Information Superhighway is a bold attempt to unify government data and information delivery state-wide. This approach is yet to develop links to community-based initiatives, but the technological capability of the system is strong and could be used as a model to link remote communities to resource management and research information in similar regions.

6.2 BUILDING INFORMATION INTEGRATION

6.2.1 Bringing together disparate information

There is the need to integrate social, behavioural and economic data into river basin management. The data need to reflect the social conditions and cultural context of resource management issues. Integrating social, behavioural and economic data with biophysical data is difficult, and the most widely practised approach, where it does occur, is the use of GIS-based decision support systems as data integration systems. These systems incorporate economic information into the analysis of decisions regarding landuse impacts, landuse changes and similar resource management processes.

> Data and information integration is rarely practised, particularly the inclusion of socio-economic information such as rates of adoption of best practices, willingness and ability to participate in catchment management, economic drivers for land use change.

CARES, University of Missouri-Columbia, developed integration procedures for multi-objective decision-making analysis. CARES built WAMADSS, a knowledge-based computer system that integrates data, information, physical simulation and economic analysis to identify alternative landuse maps for solving specific river basin problems. The objectives of the research were to: (1) develop a user-friendly, interactive watershed management decision support system that identifies the relative contribution of sub-basin areas to agricultural non-point source pollution and evaluates the effects of alternative landuse/management activities and practices (LUMAPs) on-farm income, soil erosion and surface water quality at the river basin scale and (2) demonstrate the utility of WAMADSS in identifying and evaluating landuse maps for controlling soil erosion and surface and ground water pollution sample watersheds (www.cares.missouri.edu).

WAMADSS demonstrates how farm income variables can be incorporated into systems that can predict alternative landuse scenarios and their impacts. WAMADSS could be extended further by incorporating social data such as adoption and compliance rates.

There are other opportunities to integrate social, economic and behavioural information into river basin information management systems. These include:

- Descriptions of the resource capital, which includes social, human and cultural resources.
- Use of psycho-social data in decision support systems, which include perception data, attitudes, beliefs and values in explanatory/causality models.
- Use of general public opinions and representative stakeholder group beliefs, which can be incorporated into the selection of resource management options, the testing and evaluation of options and the formulation of river basin management plans by workshop-based modelling procedures.

6.2.2 Effective river basin information management

It is now common practice for government agencies charged with natural resource management to provide river basin data and information as contained in reports (such as hydrologic data, digital elevation models, water use statistics and state of the environment reports. This information needs to be written clearly and be accessible. Examples include the State of Indiana, which has developed a searchable Volunteer Water Quality, Monitoring Database (www.ai.org/dnr/soilcons/riverbank/riverbank.html (accessed December 2004) and Wisconsin's Nonpoint Source Pollution Abatement Program www.dnr.state.wi.us/eq/wq/nps/nps2.htm (accessed December 2004)). Environment Canada produced graphic materials in paper format for environmental management of the St. Lawrence Region in the Environmental Atlas of the St. Lawrence and the State of the Environment Report of the St. Lawrence.

> Good river basin information is simple, clear, unbiased, and accessible. Visually appealing, unambiguous information ensures better understandings. It communicates the current state of the catchment and where remediation is required.

There has been considerable effort to translate research into meaningful information for a broader group of river basin managers in the Pacific Northwest of the USA. The Willapa Alliance group (a community organisation) produced a simple yet profound study, Willapa Indicators for Sustainable Community, which describes the ecological, economic and social conditions of ecosystems and communities in the Willapa Basin. Their study is published as a succinct booklet and includes indicator data on Environment (Water Resources, Landuse, Species Viability), Economy (Productivity, Opportunity, Diversity, Equity) and Community (Life-long Learning, Health, Citizenship, Stewardship).

The Willapa Alliance used indicators as directives for further research and linked with researchers of the Pacific Northwest at the University of Washington, Oregon State University, and elsewhere. The purpose was to facilitate an ongoing research programme to understand the knowledge gaps they have identified.

6.2.3 An adaptive approach—the open-access information system

Messages from research programmes are frequently complex and perhaps even contradictory. Access to data is difficult and high-tech systems may preclude community use. To overcome these issues, researchers at the University of Waterloo, Canada, developed an innovative approach to information exchange. Professor C. Mayfield (University of Waterloo) maintains:

> One of the major stumbling blocks to any environmental initiative is the lack of access to information and data in a timely and efficient manner for use by elected bodies, planners and administrators, business people, and community groups. About 80% of all data can be referenced to its geographic location (geo-referenced) and when this is available, often makes the data more useful and valuable;

and further:

> Certain natural resource management issues are suitable for expert systems, rules-based technology where these are the criteria, these are the decision points, and these are the results. But when you have something as complex and interactive as a river basin analysis, that approach

> doesn't work very well, with too many variables, too many factors, and too many decisions made not on yes/no or numerical problems but on complex legal problems, planning decisions, political decisions. The difficulty comes in how you take data and information and put it in a management plan, how you get the data in the first place and use it.
>
> (Personal communication, November 1997)

To overcome these problems, Mayfield developed an open access information system. The idea is that planners, the public and all affected participants who have a stake in getting information out should be able to have easy access. They need a common data set to find information. Mayfield developed a community information system whereby scientists and all players put information into a common database. All players buy into the concept though cash or in-kind contributions. As a rule, the system avoids issues of data provenance and reliability factors—these are self-imposed as people buy into the system into a common planning regime.

An adaptive approach to the development of information delivery system is effective.

Mayfield cited an example based on the issue of urban subdivision in southern Ontario, which threatened important cultural heritage and ecological values. There were significant community concerns being raised about the development. Mayfield maintained that the public needed to buy into the land information because they wanted a stake in what the planners were doing. There was a tendency for planners to talk *at* the public and the public to disbelieve them. Mayfield developed a solution to this problem by creating a community information system that is accessible to ordinary people, allowing them to contribute information. They used the local high schools as places where much of the information system could be handled. Students and teachers were used to set up the system and input data. Other information was added by his research team. The information system included land tax information, all laws pertaining to planning, planning acts, previous decisions, river basin management plans, Local Government planning areas with their zoning codes, all maps of the area, ecological zones, soil types and other natural features and cultural features of historic value. The system produced a cultural history and an ecological description of the village under development pressure. The information system was taken to the local council who took the detailed information and community support and rejected the development application. The system will eventually include access to biophysical information, will be made available on the Internet, and will most likely be linked to a community Intranet being set up for this region of Ontario.

The value of this approach is that it allowed landowners and planners to know what their options were. It incorporated all ecological zoning, eco-zones, bioregions, and an overlay of significant species in a presentation mode. Landuse planning consultants can use the system as it presents constraints and opportunities for land uses. References to planning acts, laws of Ontario, watershed management plans and updates to the data can be sent to developers and consultants. The information system does not propose solutions. The intention is to develop the information management system further to provide this capability, and an Internet delivery process using a structured query language (SQL) database. This is proposed for the Laurel Creek watershed (www.bordeaux.uwaterloo.ca). Similar river basin information systems are under construction for larger watersheds, such as the Grand River (www.grandriver.ca).

6.2.4 The importance of jurisdictional boundaries

Well-informed planning is an accepted principle for sustainable development in Canada, but the use of integrated information management systems is constrained by jurisdictional boundaries. In the previous section, we saw that sub-basin planning in Ontario influenced by well-informed stakeholders who obtained and used well-presented resource information. However, little cross-jurisdictional cooperation exists for information exchange at the government level. Conservation authorities, Municipal Governments and State Government agencies have yet to develop integrated information management systems that span these three jurisdictions, but where new GIS applications have been developed, they are being rapidly adopted, driven by committed technical specialists. One example is the GIS capability of the Grand River Conservation Authority, which has overcome some of these jurisdictional boundaries by developing a river basin information system (www.grandriver.on.ca).

There is growing interest in using this facility by local authorities in the Grand River basin using common GIS products. This approach has broader application and demonstrates the ability of a regional authority to transcend jurisdictional boundaries.

6.2.5 River basins as decision regions

River basins may be an inappropriate unit for information delivery. Other options could be a region of common concern or a decision-making region, defined according to the sphere of influence of a civic institution (for example, a government authority or a river basin organisation). Interplay of these three 'regions' could facilitate information delivery by paralleling the dominant locus of information flows in a region.

> The river basin may not be the most effective delivery unit—focus on the decision-making regime.

An appropriate delivery unit is one that is determined by an overlay of the region in which the biophysical causes of a natural resources management problem are located, the economic region in which resource use occurs and the sphere of influence of decision-making relevant to the solution of that problem. The St. Lawrence River Seaway economic region is an example of this phenomenon, where a natural resource management programme for water quality was built not around a biophysical region but a region of economic influence.

Two issues are important to the delivery of information at the river basin scale: what is an appropriate organisational arrangement and what are the appropriate institutional arrangements that will facilitate R&D information delivery? The answer to these questions is found in the analysis of current practices in decision-making in river basin and information management.

Information delivery to river basin management appears to be made easier when river basin management planning is facilitated and this is best done using RBOs, such as conservation authorities. These organisations have regional jurisdiction and can provide research data and information to Local Government, which is often unable to afford information management systems. In Ontario, the delivery of natural resources information to decision-makers at the river basin scale was further enhanced when centralised State Government information providers worked collaboratively with conservation authorities to provide region specific data. This procedure is still emerging in Ontario, and is facilitated by linking administrative programmes between organisations, and the presence of strong advocates for data integration operating in all organisations.

6.2.6 Involvement and ownership

Most river basin information management systems are not stakeholder driven. Where river basin information management systems exist, they tend to be set up by government agencies or derive from agency-led river basin management programmes. Information management systems which are community owned and driven produce results, such as changes to local resource use planning and policy. The Laurel Creek information system discussed earlier is a good example. This suggests stakeholder information ownership is a powerful force in river basin management.

The Grand River and Laurel Creek watershed information systems cited earlier serve as models for cost-sharing and joint ownership. More effort is needed, however, to identify the specific organisational arrangements, the ownership of intellectual property rights, accessibility issues and technology requirements if this approach is to have widespread application.

> End user involvement to design and jointly owned, cost-shared processes facilitate information delivery.

6.2.7 Visualisation

Visualising river basin information is a powerful mechanism in changing perceptions of resource use and increasing the likelihood of using research results. (Malczewski, Pazner and Zaliwska 1997) demonstrated the importance of cartographic design in increasing the ability of end-users to interpret the results of multi-criteria location analysis. They maintained that GIS is a powerful tool for aiding decision-making, not just in the storage, retrieval and manipulation of data, but in visualising the results of analysis. They note that quality of graphical presentation and the way of conveying information to the decision-maker might significantly affect the decision-making process. This comment should be kept in mind in the rush to use web-based information systems, which provide a nearly infinitesimal array of visual possibilities.

Significant visualisation of river basin R&D data is now widely practised such as at:

- www.cares.missouri.edu for examples of floodplain, small watershed and farm visualisations, and
- www.cgia.cgia.state.nc.us for state-wide data visualisations of North Carolina by the Center for Geographic Information and Analysis.

6.2.8 Issue-driven solutions

A common feature of the information delivery systems, data integration processes, modelling techniques and publicity programmes is that critical, highly emotive, 'hot' issues drive the need for and the processes of information delivery. Political processes and community outcry are frequently the catalysts for action. Researchers and government agents respond to these demands and provide solutions to a concerned public.

> Effective information delivery, like catchment management, is driven by hot issues and innovative institutions.

This issue-driven approach suggests that information delivery for river basin management is not really a technical issue (how to construct GIS, develop explanatory models, develop workable data transfer systems and have the best equipment) but rather an institutional/organisational issue. It is not that technical issues are irrelevant. On the contrary, appropriate, user-friendly technology is critical if effective use is to be made of river basin information. Rather, it should be acknowledged that political processes drive river basin management, and the failure to deliver river basin information is not due to the lack of technical expertise, but rather due to the failures of governments, business, communities and citizens to work together. Institutional issues such as the governance of the river basin (who is 'in control'), institutional arrangements (taxes, incentives, regulations, cost-sharing arrangements) and organisational structures for river basin management (types, roles and responsibilities) are significant factors in delivering river basin information.

6.3 A PROTOTYPE RIVER BASIN INFORMATION EXCHANGE PROGRAMME (RBIEP)

6.3.1 Components

An information exchange programme for river basin management provides improved information exchange to support basin governance. This was previously reported (Hooper 2002) and here a summary is presented. The RBIEP has several components.

(1) Interactive community education programme
This component is the main driver for river basin education aimed at the residents of the river basin, business, researchers and State Government operating within and outside the river basin. The aim of the programme is to support IRBM planning by providing similar messages in all current education and media programmes and avoid conflicting advice. The education programme integrates current educational activities and links these to the best management options available in the River Basin Management Information System (RBIS)—items 7–9 below. The education programme also includes specific courses on river basin management methods tailored to suit regional needs.

(2) Community-based river basin information centre
This is a facility in the river basin which contains information collections such as atlases and indexes; operates a farmer phone-in service, a fax server, a mailing list, a 1-800-river basin phone link, websites and email list servers; undertakes community education programmes; develops and implements conservation partnership agreements. The River Basin Information Centre is housed in the headquarters of the river basin management organisation and reports to that organisation.

(3) 1-800-river basin phone link
Knowing that the spoken word is a powerful medium for information exchange, a 1800 service provides fast service on natural resources management. There may also be people with limited reading and writing skills, or with limited Internet capability who prefer to exchange information orally. Business managers may want to know information quickly and may be unwilling to spend a lot of time in formal education or training programmes. What they want is a quick answer about a resource management question from a river basin management organisation.

The 1-800 phone link, provided by the river basin management organisation, is an inquiry service. It provides advice on the development and implementation of land and water management plans, and cost-sharing arrangements for best management options for farmers, river basin management organisations and governments.

(4) A charter on the legalities of information sharing
This is a set of information management protocols which address legal liability, reliability, ownership, copyright and freedom of information issues. These are presented as Management Guidelines which describe the legal obligations of users. These protocols include the wording of website disclaimers.

(5) Institutional arrangements for river basin information exchange programme (RBIEP) implementation
Effective river basin management is more likely to occur when river basin management organisations are stable, well-funded, have well developed corporate plans and business plans and are cost-shared between government and regional

communities (AACM International & Centre for Water Policy Research 1995). Studies in Australian river basins have shown that river basin managers prefer a strengthening of the voluntary approach used in river basin management today (Hooper 1999). River basin managers prefer an organisation which has a mix of the powers of an authority and the functions of a tribunal, yet without losing the 'grass-roots' enthusiasm of voluntary approaches. Similarly, river basin information management needs to be managed by a stable river basin management organisation to be effective. The RBIEP is best established in a stable organisational arrangement in which the cost of implementation and maintenance is cost-shared amongst stakeholders.

(6) Regional libraries—safe repositories of river basin information
University libraries offer river basin managers the advantage of being stable (long term funded) institutions. They frequently have special regional collections ideally suited for keeping material for river basin management (like reports, strategies, plans) in a climatically controlled environment, in which there is commitment to document preservation. Librarians have strong cataloguing skills and can provide river basin management organisations with the logical organisation of material.

University libraries are also research repositories, committed to collecting and preserving high quality international publications as well as home-published anecdotal information. For these reasons university libraries are part of a RBIEP. They are the place where stakeholders can locate hard copies of river basin documents. This can be done simply and quickly by remote electronic access by individuals or river basin management organisations, using a sign-up contract.

(7) River Basin Information System (RBIS)—interactive webpages and searchable databases
The RBIS has a number of Webpages and relational databases shared through open and restricted access websites and reproduced on a CD-ROM. The structure of Webpages in a RBIS is shown in Figure 6.1. The Land Types and Best Management Options database is the most important RBIS feature in that it lists the best management options for each land unit at the sub-basin level (Figure 6.2 shows a prototype design). The RBIS also contains a *Join Discussion* facility to provide instant and ongoing feedback to river basin management stakeholders about the best management options. The feedback is catalogued in the same database and is searchable on the web. This database provides a growing archive of comments on river basin management options.

The RBIS includes a search capability on river basin documents, research projects and management skills in the river basin, allied to each best management option and according to land unit and sub-basin. Thus, an inquirer can easily find the best management option, supporting research and key contact person to discuss the validity of that option. The database can be built to include multi-media data (voice, graphics, movies, photos, numbers and text). A separate, restricted access database is available for the mailing list of the river basin management organisation. The entire set of databases and Webpages

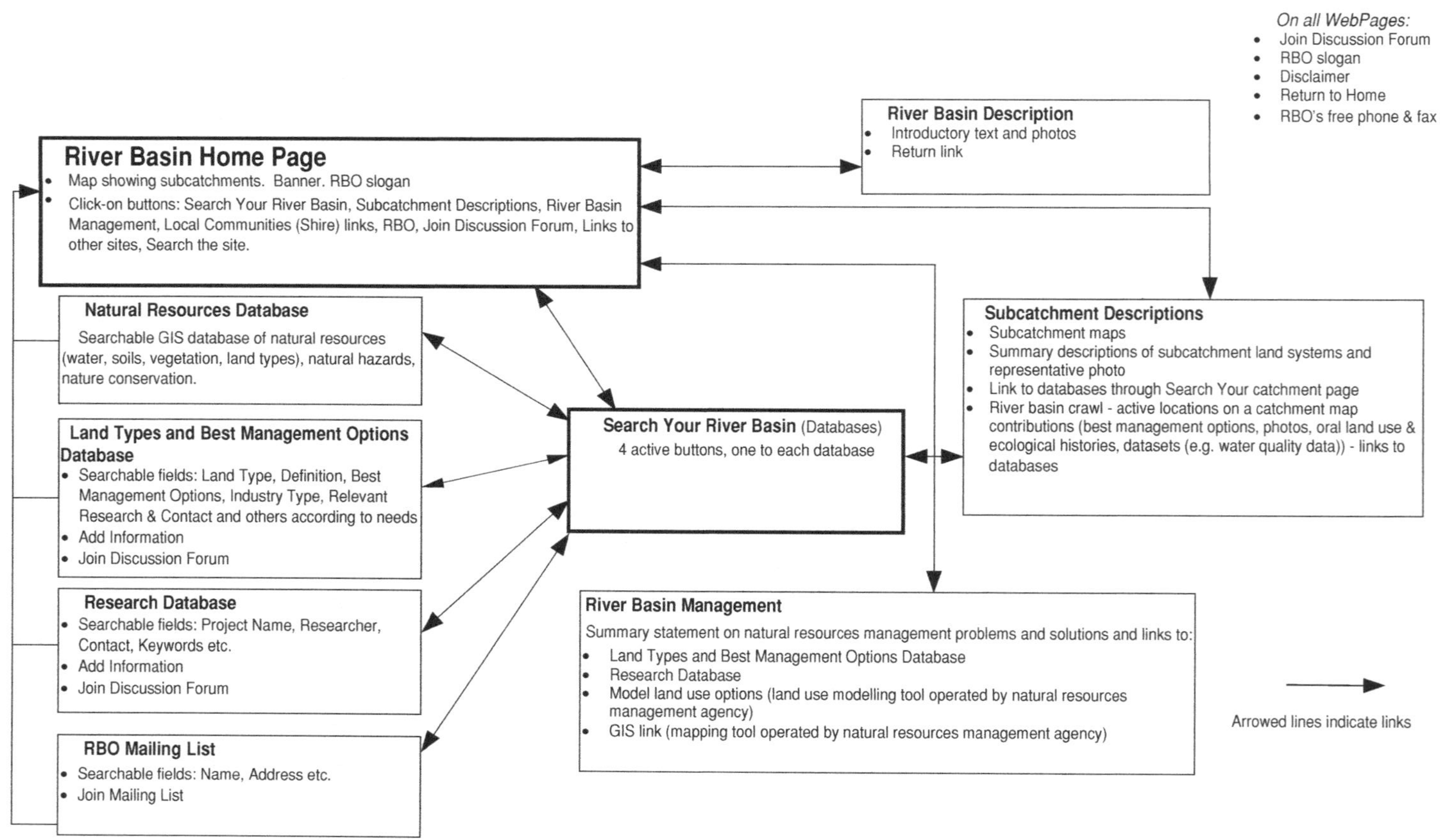

Figure 6.1 River Basin Management Information System—webpage structure.

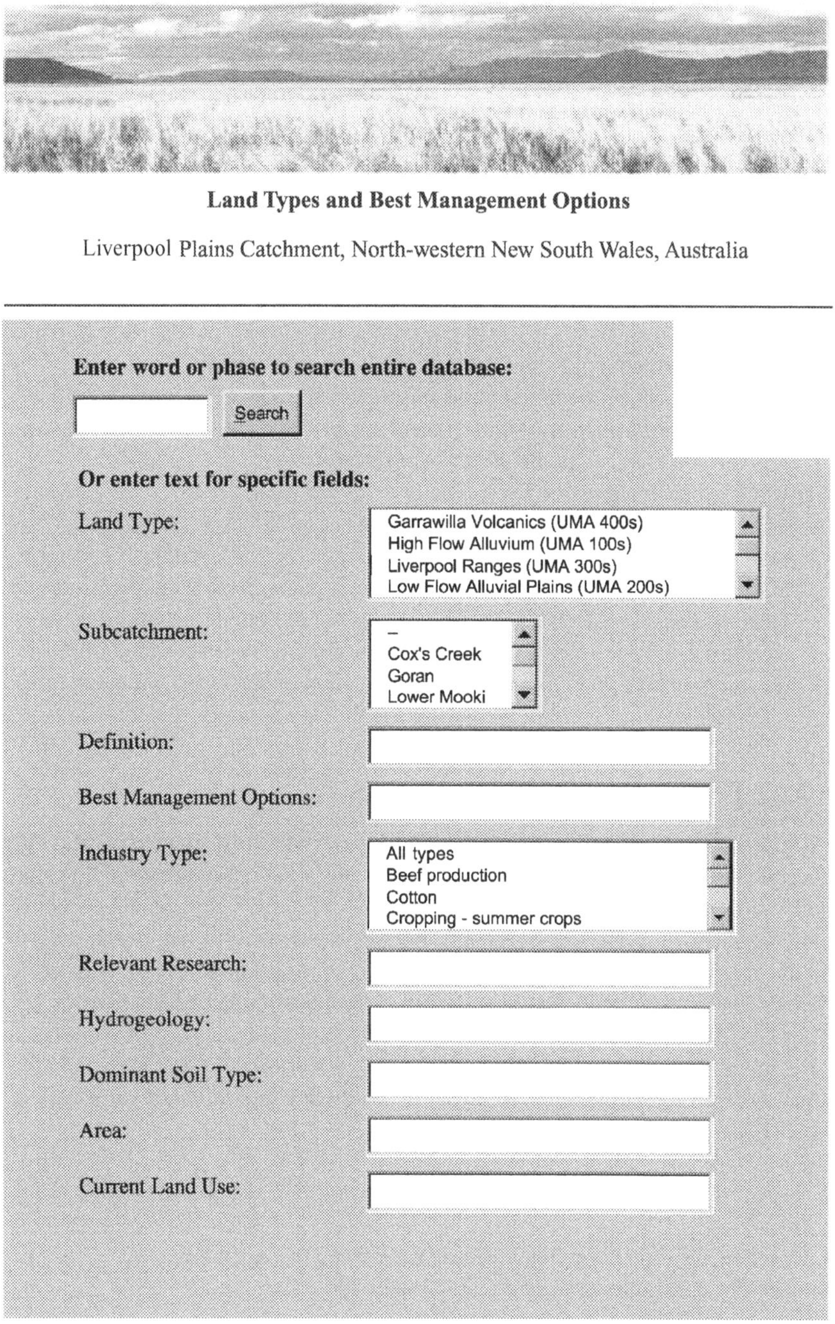

Figure 6.2 Land types and best management options database search page in a prototype RBIS.
Source: Author.

can be built as a database. This allows the RBIS to be transportable from river basin to river basin—allowing river basin management organisations to design an RBIS relevant to their needs.

(8) RBIS—modelling and decision support
This feature of a RBIS is a multi-objective decision-making tool, which provides planning support. It is also an educational tool which presents resource use options to river basin managers. The decision support system presents land management options on single land units, and incorporates social and economic data and biophysical models. It also interrogates and models resource use practices which are nominated by river basin managers (farmers, agri-business managers, Landcare leaders, State Government agency staff and Local Government planners) in workshops, or individually. The results of the modelling can be shown on a point and click map on a webpage as a searchable geographic information system. Alternatively, a user can download modelling tools which then run on his/her own computer.

(9) RBIS—a searchable GIS
This is an interrogative electronic atlas of the river basin. It allows the user to search land types, landuse, best management options, settlement patterns, land ownership and planning zones in a river basin. It is designed for use by all river basin mangers, a user-friendly system on the Internet and it includes a river basin visualisation technology to display the results of modelling and decision support systems, including the location of best management options.

6.3.2 Preferred contents of an RBIS

Hooper (2001) found that in open-ended interviews with river basin managers that many commented that river basin management organisations use information similarly to the method used by a board of directors or heads of government agencies: they do not require basic information, rather processed information that allows them to make effective decisions. They operate with different levels of scientific literacy and operate under time constraints. There is no time to access databases. They require information prepared for them to be filtered so that they can access core, interpreted information. This implies that a river basin information management system requires people employed to synthesise and interpret information. The RBIS:

- provides biophysical, social and economic and landuse characteristics of the river basin,
- specifies roles and responsibilities in river basin management,
- specifies the results of research in common language,
- is designed around issues and localities which could inform river basin management at that level,
- includes analysed information, not raw research output nor raw data, preferably in a verbal form as a presentation to a river basin management organisation,

supported by a paper copy, and interpreted within a river basin management context,

- includes options for river basin management about which stakeholders make decisions. The information must be accurate and reliable to allow for the quantification of the options.

Nine types of information in a RBIS were preferred by the river basin managers in a survey of the information behaviour which accompanied the development of the RBIS (Hooper 2001) and are listed in Table 6.1.

Table 6.1 Preferred information types in an information system (in rank order of response) $n = 73$

Category	Examples	Frequency of responses
Natural and cultural resource information	Biophysical data on soils and water resources, socio-economic characteristics of river basin, historical trends in river basin data	52
River basin management	Responsibilities, legislation, issue locations, farmer rights, contacts, networks, project evaluations, success stories	38
Research	Results and evaluations of research projects, assumptions in research, researcher contacts, decision support models, Internet sites, technical information	28
Landcare initiatives	Contacts for Landcare groups, conferences, field days, newsletters, examples of success stories at local level	16
Best management options and guidelines for natural resource management	Problem solving techniques, evaluations of best management options, linking issues to solutions and evaluating practice	13
Strategic planning in natural resources management	Strategic plans, long-term funding processes, framework for regional strategies	11
Production and infrastructure information	Irrigation development, market research, productivity changes	8
Community input	Evidence and provision of community input facility, access for children to information system, use of anecdotal information	8
Environmental reporting	State of the environment report for river basin, evidence of cumulative impacts, performance indicators	5

The interviewees in the study also recommended the following characteristics of an information management system:

- be able to continuously updated,
- allow for community involvement and community input,
- use CD-ROMs and be accessible on the Internet,
- be used as a community consultation technique,
- be built around a river basin home Website,
- be searchable,
- use visual aids (graphs and tables) accompanied by explanations,
- be up to date, simple to access and use,
- be either a newsletter, a newspaper, or an electronic media or all three.

Table 6.2 shows typical questions the river basin information system can answer.

6.3.3 The broader information exchange programme

The prototype RBIEP described above was developed in response to local needs. It involved integrating information from many sources, providing it to decision-makers in a simple, straightforward way, and providing river basin management stakeholders with a facility to easily share their ideas about the information. It involved facilitating an Internet-based system of information exchange available to such stakeholders as: farmers, Landcare coordinators, community groups, business leaders, researchers, river basin management groups and government decision-makers. The RBIEP places importance on all relevant information sources, not just 'technical experts'. It engages local communities by providing opportunities for discussions and simplifying the ability to provide and access information. Most of the information in the programme is stored in the RBIS component of the RBIEP.

The RBIEP is implemented using a training programme with river basin managers based on the User Guide, a self-guided evaluation of information management for river basin management with chapters on each of the above components, and was published on the Internet at www.catchment.com/CIShome.html. Each component of the RBIEP in the User Guide has Key Questions, which allow Guide users to self-assess their needs in each of the components.

Why use the Internet?

There are many different types of people and organisations in river basin management—stakeholders with different needs and different backgrounds, different decisions to be made at varying scales and locations. One way to bring these stakeholders together is to use the ubiquitous Internet-linked computer. The Internet offers new possibilities to assist word-of-mouth information exchange. It does not replace it, but helps stakeholders to share information over longer distances. The Internet can also be used to order information from libraries and receive it electronically or in the post. In addition, the Internet can be used as a virtual library.

Table 6.2 Examples of river basin information system functions

Functions	Question or enquiry
Join discussion	
Contacts	Where can I get information on best management options for different types of land in our catchment? Who knows the correct type of tree species to replant for shelterbelts at location 'A'? Who can I call? What is the current water allocation plan in the catchment? How does it affect my irrigation plans at site 'B'?
Feedback	What were the community's responses to dam building proposals in the valley? What do farmers think about this best management option?
Guidelines	
Guidelines for best management practices	List best management options for each land type in the sub-catchment and what are the relevant planning policies of Local Government? What is the research to back up the best management options? Who did the research? How do I contact them?
Guidelines for cost-sharing agreements	Outline the cost-sharing agreements for the River basin Management Plan Identify benefits for my catchment. Who can participate? Please send me the application form for cost-sharing
Local knowledge	
Local knowledge	What is the history of farming in this catchment? How has the ecology of this area changed over time? What are peoples' opinions of the best management options at point 'C'? Who has tried a best management option and has it worked?
Where is?	
Geographical Information System and visualisation (electronic atlas)	What are the soil and water resources in my sub-catchment and where are they located? Where are the pristine river habitats in the river basin? Where is land type 'D' found in the catchment? Fly me over the river basin using a 50-year prediction of a 100%, 75%, 50%, 25% adoption of vegetation re-establishment under Planning Policy 'E'
Who are we? What do we do?	
Community profiles	Describe the community of the valley (population size, age structure, income distribution, household income, housing, age profile, employment levels). What are the economic impacts on the community profile of different best management options?

(*Continued*)

Table 6.2 Examples of river basin information system functions (*Continued*)

Functions	Question or enquiry
Adoption rates	List current, past and intended adoption rates of best management options for soil conservation
Business directory	Identify all businesses with an interest in wheat production, and show me their area of influence. Where do I go to get information on soil moisture probes?
River basin management	
Planning constraints	Find all land of type 'F' and list the planning constraints on that land type
Model landuse options	What are the economic and environmental impacts of zero till cultivation on a groundwater recharge in sub-catchment 'G'? List the potential and predicted impacts of dam building within 10, 50 and 100 km downstream on river 'H'. What are the groundwater impacts of a 50% increase in irrigation in this sub-catchment?

The Internet is not the solution to river basin management information problems. The Internet can provide people with a facility to assist the process of word of mouth information exchange, but never replace it. There are also problems with the use of the Internet: copyright, privacy, slander and information overload. Many rural communities have limited Internet access. For this reason a 1-800 (free call) phone number was recommended to link resource managers to the river basin management organisation where an advisor can handle their inquiry, search the databases and respond. If the information is not available immediately, a fax or surface mail is sent when the information is found.

Best management options—the first foundation of the RBIS

The design of the RBIS was based on two fundamentals. The first was the concept of 'best management options', which refers to the most technically feasible, socially relevant and financially viable management practices to manage land and water resources at a specific location in a river basin. These options are usually specified in the context of an overall River Basin Management Plan and state natural resource management and economic consequences to both the resource manager and the river basin community using the options. This allows decision-makers to establish their preferred options based on prescribed criteria and sound technical, social and economic judgements. River basin management organisations have an important role to play in identifying best management options. They do this by accessing and reviewing research outcomes, and providing specific information about resource data, information and options as provided by the river basin information system.

Land units—the second foundation of the river basin management information system

The second foundation for the design of the RBIS was the concept of a 'land unit'. A land unit is an area of land best thought of as an integrated ecological unit. Land units are unique areas of land defined according to a combination of geological material, soil, slope, vegetation and water resources, and over which there is a common set of constraints and opportunities for landuse (Christian and Stewart 1968). The biophysical features of a land unit determine these landuse possibilities. Best management options can be identified for specific land units in a river basin.

Land units can be any size but they 'aggregate up' to form land systems within a river basin. Most river basins comprise many different land systems and land units. This means different possibilities for landuse exists in different parts of a river basin according to the characteristics of the land, technology and capital availability. Consequently, each land unit can be defined in terms of suitable land management options.

These two fundamentals were used in the design of the Land Types and Best Management Options Database in the RBIS. This is discussed in RBIEP component (7) above.

6.3.4 Implementing the prototype RBIEP

The prototype RBIS was tested at the two field sites. This involved using a Project Reference Group to practise using the databases and completing a questionnaire on its usefulness to their river basin management work. Workshops with two Projects Reference Groups in Australia were used to discuss the content, structure and usefulness of the prototype. Adjustments were made to the prototype databases and the following protocol for implementing the RBIEP was developed. As well, an Australia-wide workshop programme was undertaken to promote the RBIEP throughout Australia in 2000, with over 300 people attending. The prototype was also published as a resource on the World Wide Web (www.catchment.com/CIShome) (on October 16, 2002). Recommendations were also made to National Government on how to implement the prototype as a national programme to enhance river basin management information exchange in Australia.

The cost of implementing a RBIEP was estimated to be approximately US$ 40,000 per river basin, comprising US$ 30,000 staffing costs (information manager), and US$ 10,000 for software and hardware development and training workshops. In this costing, the RBIEP was owned and managed by a cost-sharing arrangement between the RBO, Local and State Government, local land and water management groups and private sector interests.

Principles

As a result of the national and project workshops discussed above, the following principles were developed and are recommended when implementing a RBIEP in other locations:

(1) An RBIEP should not attempt to replace, but enhance, word-of-mouth networks, conferences, telephone conversations, newsletters and booklets. The RBIEP should be the repository of river basin information, a central place of information exchange, directing inquirers to hard copies of river basin material, which are stored in regional and local libraries.
(2) It is vital to employ an information manager in a river basin to manage the RBIEP. The person should be employed under an agreement between Local Government, State Government, private sector interests and the river basin management organisation in the river basin.
(3) Information for river basin management decision-making must provide advice for the development of guidelines and requirements for resource use, expressed as best management options, so as to meet policy and plan requirements.
(4) River basin information should be assembled so as to address specific issues in specific locations, attached to prescriptive land management units at the sub-basin level and related to best management options. Sub-basin solutions must be developed within a whole river basin context and information for a sub-basin must recognise broader river basin-wide implications.
(5) An RBIS should not duplicate metadata and information systems and extension programmes by State agencies. A RBIS could make government systems more useful, by being a first point of contact in a river basin.
(6) There is the need to use current and emerging electronic communication networks in rural locations to house the RBIS.
(7) An RBIS should be embedded in the river basin management decision-making procedures, using this approach:

- river basin management bodies identify issues,
- the same bodies commission ongoing research to identify options and analyse consequences,
- the outcomes of each of the options to address the issues are quantified and qualified, specific to sub-basin locations,
- a community engagement process is used by the river basin management bodies to present these options for discussion and endorsement throughout the river basin,
- river basin management bodies present these options in the RBIS.

Outcomes and performance

The project to develop a RBIEP and its component RBIS was fairly successful in that it became owned by the stakeholders in both river basins. It has been selected by the Australian Government (which provided the majority of funding of the study to build the prototype) as a project to be tagged for a ten year monitoring of project outcomes. There was also endorsement of the RBIEP by the funding agency and selection of the RBIEP as a member of Land and Water Australia's Innovations Database in natural resources management research and development (http://www.infoscan.com.au/id/web/browse.htm, accessed, December 2004). At the international scale, the RBIEP approach was selected as a case study in information exchange for the Global Water Partnership's Integrated Water Resources Management ToolBox (www.gwpforum.org).

The project raised awareness of the need for improved exchange and integration at the national level, but it has not been fully implemented. A national RBIEP programme, recommended in the project reports to the funding agency, was not established in Australia. This suggests that considerable work is needed in capacity building in IRBM to build the institutions capable of implementing a RBIEP.

Furthermore, an RBIEP will not be sustainable in the long term unless there is significant investment by governments in the project products. There is significant public good benefit (in improved adoption of best management practices in natural resources management) which should qualify the RBIEP for public financial support, and an independent consultancy review demonstrated a significant cost benefit to the field site river basins if the RBIS was implemented.

Problems encountered during the implementation phase

The key implementation issue for a RBIEP is that river basin management organisations in Australia do not have secure institutional arrangements for their existence. This means that they do not have a strong enough funding base to invest in long term information exchange and brokerage amongst their stakeholders. The same weaknesses which made it difficult for them to manage river basins effectively (lack of jurisdiction and secure funding) also make it difficult to manage river basin information effectively amongst their own stakeholders.

Another implementation problem is the reluctance of governments to exchange information sets they possess freely on the Internet. The concerns are more than copyright issues, and include fear of litigation, misuse of data and concern the information they provide will be used for commercial purposes. These problems have not been resolved and will require further innovative institutional arrangements in order to secure more rigorous river basin management entities and information exchange protocols by and between government agencies and community organisations.

6.3.5 Lessons learned and replicability

The RBIEP is a unique programme in that it can be co-owned and managed by government and community-based river basin management organisations. There was sustained enthusiasm by the river basin management communities engaged in this project to build a prototype RBIEP. The critical issue is to establish a long-term sustained funding mechanism in place for river basin management organisations.

However, a number of institutional and social issues have precluded immediate implementation of the prototype such as:

- lack of coordination mechanisms to share information,
- lack of short-term funding of river basin management organisations,
- lack of leadership in government and river basin management organisations to build regional information exchange protocols and processes.

Furthermore, issues such as the lack of data sharing protocols, the scepticism in government about farmers' best practices, loss of scientific ownership of project results by researchers, and an unwillingness to handle co-managed, co-owned information management systems have proved obstacles.

These are difficult problems and need to be addressed. Consequently, a set of institutional arrangements to facilitate implementation were developed in the RBIEP. These included a cost-sharing arrangement between State Government agencies, Local Government entities and river basin organisations. The focus for the RBIEP is a local planning application, in which planning instruments are used as river basin management tools. There are also opportunities to expand the type of information in a RBIEP to include more general social and economic indicators. This may facilitate more 'buy in' by social welfare and planning departments to support the formation of the RBIEP. In this case it might be possible to create a whole of government information service at river basin level. For example, community groups in social welfare situations may identify benefit in attaching some of their activities to whole of river basin management programmes.

Despite the difficulties outlined above, the RBIEP has high potential for replicability. The institutional arrangements, which can facilitate adoption of the RBIEP and the design of the RBIS on the World Wide Web are easily transferred to other contexts, although they would require adaptation to the unique conditions of each site.

6.4 REFERENCES

AACM International & Centre for Water Policy Research. 1995, Enhancing the Effectiveness of Catchment Management Planning. Final Report. Final Report for the Department of Primary Industries and Energy. Adelaide, AACM International Pty. Ltd.

Christian, C. S. & Stewart, G. A. 1968, Methodology of Integrated Surveys. Aerial Surveys and Integrated Studies. Proceedings of the Toulouse Conference. UNESCO. 1964. 233–280.

Hooper, B. P. 1998, Delivering R&D to Catchment Managers: A Review of Selected North American Experiences and Implications for Australian Practice. UQL15 Research and Development Final Report (Land and Water Resources Research and Development Corporation). Brisbane, University of Queensland.

Hooper, B. P. 1999, An Integrated Information Management System for Catchment Managers. UQL10. Research and Development Final Report (Land and Water Resources Research and Development Corporation). Integrated Resource Management Research Pty. Ltd.

Hooper, B. P. 2001, Information management behavior of Australian catchment managers. Water International 26[4], 513–526.

Hooper, B. P. 2002, Design and implementation experience of an Australian prototype catchment information exchange program. Water International 27[4], 568–577.

Malczewski, J., Pazner, M., & Zaliwska, M. 1997, Visualization of multicriteria location analysis using raster GIS: A case study. Cartography and Geographic Information Systems 24[2], 80–90.

7

Social dimensions, institutional arrangements and performance measurement for IRBM

7.1 THE POWER OF THE SOCIAL DOMAIN IN IRBM

7.1.1 IRBM—a problem scenario

This chapter focuses on how the social domain and improved institutions provide the power to make coordinated decision-making possible. It is axiomatic that more biophysical research will lead to improved understanding of river basin processes. Accurate biophysical information will inform river basin management decision-making, and improve the ability to make effective natural resources management decisions. There is greater understanding of biophysical processes today than ever before and the question is do we have enough knowledge and wisdom (as discussed in Chapter 5) to implement IRBM? Perhaps, but the contention of this chapter is that what is missing is a strong knowledge of the social and institutional domains. Knowing these will provide more effective basin management decisions.

The problems of basin management are complex and locked in both spatial and temporal settings. When discussing river basin development projects and management, Barrow (1998) stated that,

> It is difficult to overcome the problems of complexity and find a practical framework to better integrate environmental, socioeconomic and policy issues at a regional scale.
>
> (p. 180).

 ISBN: 1 84339 088 4.

In Chapter 2 (Table 2.3), the following complexity-related problems were listed: inflexible planning, lack of data, poor assumptions, over reliance on one-dimensional solutions, development which precludes local ownership, negative impacts of development projects, differing developing/developed country priorities, lack of an RBO power base, lack of RBO accountability, failure to manage whole basins, border conflicts at the basin edge, weak coordination of management instruments, political influences, problems of multi-lateral jurisdiction and sectoral dominance.

These problems clearly indicate that social and institutional issues are the critical impediments to effective integrated river basin management. The failure to achieve effective results is constrained by factors such as:

- Ignorance of the range of institutional arrangements that facilitate integrated management techniques.
- Lack of ongoing funding.
- The need for a river basin advocate who can transcend local interests, overcome government inertia and champion the need for a regional approach to integrated resource and environmental management.
- Leadership skills in basin management organisations.
- Lack of knowledge of the decision systems in river basins.
- Lack of effective legislation to enact basin management, or where it exists, lack of implementation.
- The need to incorporate differing values and expectations about resource use.
- Definition of the management regime: should river basin organisations be used when water use is dominated by groundwater extraction? Should groundwater basin management organisations be developed? If so, what institutional arrangements are best for conjunctive use settings?
- Failure to use economic analysis to cost the benefits and disbenefits of management actions, and develop mechanisms to share costs both within the river basin and beyond the basin for public good benefits to society in general.
- Ignorance about the drivers of regional economies and how these impact on basin resource use.
- Too big a solution: are basin-level solutions best when only sub-basin management solutions are needed?
- Potential higher costs: does river basin management generate more administrative costs, even though it allows improved public participation?
- The lack of use of social impact assessment to determine net social gains from implementing river basin management plans.
- The lack of ongoing financial and infrastructure support by government to community-based, voluntary river basin management initiatives.
- Increasing evidence of burnout by volunteer river basin management participants and others.

Source: Adapted from Le Moigne et al. (1992), Le Moigne et al. (1994), Barrow (1998), Hooper, McDonald and Mitchell (1999), Shah, Molden and Sakthivadivel (2003b).

These factors require attention to improve river basin management. Much work has been done on the smaller scale of subregional North American watersheds to identify entry points for improved governance and, while this experience cannot be directly translated into river basins elsewhere in the world, it should be considered. Lant (2003), for example, identified five characteristics that watershed-based institutional structures must possess to be valuable to natural resources management:

- Enjoy the type of power and authority generally associated with centralised administrative governments but also establish democratically based legitimacy and regional and local levels.
- Have the authority and responsibility to manage watershed issues holistically—on a systems basis.
- Rely on more than voluntary measures (including financing mechanisms and compliance mechanisms).
- Capacity—budget, staff and expertise.
- Should be generalised across watershed types, scales and political units, with information gathering capacity and protocols for information sharing.

Similarly, Born and Genskow (1999), Born (2000), Born and Genskow (2000) and Born and Genskow (2001) encapsulated the fundamental problems and provided directions to solve complexity in watershed management. In these four comprehensive overviews, they demonstrated clearly that watershed management in North America is more likely to be successful when driven by strong institutional arrangements. Their findings could apply to river basin management but at a broader scale and one of increasing complexity:

- Importance of context—it determines what is appropriate and can be achieved—so the degree of organisational formality varies according to the setting.
- The ability of RBOs to mobilise funding/staffing—a survival strategy.
- There is no single organisational model—avoid the 'cookie cutter' approach; different formal structures and different provisions needed for different settings.
- An inclusive, diverse membership precipitates ongoing ownership.
- A sustained government role is essential.
- Adequate scientific data and analysis, sound planning processes lead to good plans and projects.
- Masking the land connection often occurs—an overemphasis on water planning erodes the use of an integrated approach.
- Integrated, ecologically sound actions happen incrementally.
- A watershed partnership authority is most likely to be the most effective entity at the sub-regional level.
- Realistic expectations—be realistic in terms of time: integrated watershed management is forever, so there is the need to think about the time it takes to produce results.
- There remains the tyranny of local landholder rights, e.g. in managing vegetation for biodiversity; so local entity prescriptions are difficult.
- There is much uncertainty in institutional interventions: there is the need to monitor, learn and adapt.

Improving governance is best done by improving the capacity of the decision-makers in a basin setting to collaborate and address the implementation constraints listed above. Imperial (2004), for example, recommended a performance management system for collaborative management in watersheds as a tool for improving organisational capacity. His approach revealed how performance management systems can be used as a motivational tool but can be fraught with difficulties.

But these questions must still be asked: is IRBM the panacea and can it improve natural resources use and management? The answer lies in recognising that IRBM provides a means to an end. That end is sustainable development and the management of scarce natural resources is to provide services to people without environmental degradation. The remainder of this chapter discusses how this is best done: by strengthening the social domain, building more robust institutions, and developing performance measures for RBOs.

7.1.2 IRBM—a dialectic, conversation-based social decision system

One of the first elements of strengthening the social domain is to recognise that IRBM is about people and organisations: how they make decisions in a basin management setting; it is about how they interact, who gets affected, who are the winners, who are the losers, how they communicate and who is involved.

In Chapter 3, decision-making was shown to be more like a 'messy' rather than an ordered, structured process. This discussion concluded by advocating an adaptive, integrated approach to river basin management decision-making, one which uses 'policy experimentation'—testing river basin management options in reality and learning from the 'doing' of those options to improve decision-making (Rogers, Roux and Biggs 2000). The approach focuses primarily on a dialectic process: a process of dialogue to engage, build partnerships, jointly identify agreed goals, corporately review results and reiterate and discuss tested solutions to improve natural resources management, and being continually able to adjust to the messy administrative, political, economic and social forces at work (McCool and Guthrie 2001).

Dialectic

—Proceeding by or as if by debate between conflicting points of view.
—Pertaining to or of the nature of logical argumentation.
—A process of change that results from an interplay between opposite tendencies.

Source: Delbridge (1982).

Others have elaborated this approach. Bruins, Heberling and Maddock (2005), for example, 'diagram' integrated management so as to explain process and include techniques such as cost–benefit analysis (CBA) combined with qualitative considerations, stated preference techniques, and linking ecological and economic models to simulate feedbacks. Born and Genskow (2001) identified that collaborative decision-making is informed by biophysical and social science, as well as local knowledge, to generate sound diagnoses of problems, clear directions and feasible actions. This 'good' planning process serves in part as a socialising processes which ground watershed management and establish joint goals and priorities for partnership activities. Falkenmark et al. (2004) demonstrated the need for effective dialogue as a precondition for effective catchment management. They strongly put this case by advocating,

> the challenge now is to create management systems where the formal decision-makers interact with relevant members of the scientific community, users and other stakeholders for a coordinated approach that successfully orchestrates water uses towards internal compatibility. (p. 297)

Dialogue

—Conversation between two or more parties.

—Especially in diplomacy, a state of communications between parties, countries, etc., in which cautious goodwill may lead to formal agreements.

Source: Delbridge (1982).

In all these examples, a common thread is *dialogue*: building understanding of each others' positions and the natural resources management problem, talking through best management options to get agreed solutions, reviewing those solutions and learning by doing this together (an adaptive process). This dialogue forms the basis for effective, collaborative decision-making in a river basin's social decision system (see box). As discussed in Section 3.6.1, the social decision system is likely to be episodic, opportunistic, complex and driven by media and political agendas than a purely rational, logical, linear process. What drives change in the social decision system is a dialogue, and at many levels.

River Basin Social Decision System—defined as the range of choice selection procedures amongst frequently competing objectives in a river basin setting operating at the basin and sub-basin (sub-regional and local) levels.

The concept of dialogue-driven decision-making has been extended significantly by Green (2004) who advocated the concept that integrated catchment management in the United Kingdom is 'conversation'. By this he meant that the ways agencies engage their stakeholders has many of the characteristics of conversation:

- It is purposive; the intention of a conversation is to change the self or others, or the relations between them.

- It involves speaking and listening; the feedback from the partner(s) being important to gauging the success of messages it is sought to convey and hence to modifying that message in order to be more successful in inducing the desired change.
- It is consequently a learning process.
- It is embedded in and expressive of social relationships which in turn define the appropriate forms of conversation.
- Thus, it is governed by rules and conventions.
- One purpose of a conversation is to build and maintain those social relationships (e.g. trust, social capital).

Green goes on to stress that conversation is more than a technocratic definition of communication in which one gives information to another; that success in conversation comes from listening to the other; that it is about building social relationships, not an attempt to change others' point of view; it is a two-way dialogue that communicates each other's position (and is interpreted as an image of each other's position) and it is a process of social learning, as a method of defining 'where we are going' (the end). In a river basin context, these observations echo the realities of anyone involved in river basin management: the key is listening and understanding each other in order to progress the agenda of IRBM.

Some pointers to improve dialogue/conversation skills for river basin settings include

- Effective dialogues are built on mutual respect and trust: this takes time—even years before stakeholders unknown to each can begin to build trusting relationships.
- In any resources management setting, there are winners and losers. Care is needed in speaking to both: to ensure equitable solutions will be sought and complaints are heard and acted on.
- Workshop leaders, programme leaders and agency resource managers frequently need specific training in conversation techniques, conflict resolution, listening skills, procedures for running meetings, how to engage high-level bureaucrats as well as common people.
- Dialogue is often a one-way street with the media and politicians: care is needed to speak clearly, and with plain language that can be understood by non-technical listeners.
- Build open dialogue processes whenever stakeholders meet: allow time for conversation so that all can have a say in meetings.
- Never underestimate the value of social interaction (coffee breaks) at meetings: this is when people do most networking and one-to-one communication.
- Emphasise the context about which you disagree (if you do) rather than focussing on the person—work towards other's points of view even if you disagree with them and respect those positions.
- Always have a listening ear and let criticism be constructive rather than putting down the listener or bolstering your own position.

7.1.3 Strengthening dialogue processes to overcome the 'tragedy of the river basin commons'

In Chapter 3 (Section 3.1.1), I suggested that a river basin operated something like a decision 'commons'. The management of this complex commons, where natural resources management problems are everyone's but no-one's problems, requires institutional arrangements where a dialogue strengthening process is put in place as the first step in building trust and confidence between stakeholders. There is a need to harness political support for river basin management simultaneously, so that a dialogue with political decision-makers supports local and basin-wide dialogues. Underpinning this approach is the need to understand the governance regimes, the political processes and the role of advocacy in the basin setting.

These governance issues were eloquently stated in the 2003 statement on integrated water resources management at the 3rd World Water Forum, where delegates agreed that,

> The key issue confronting most countries today is that of more effective governance, improved capacity and adequate financing to address the increasing challenge of satisfying human and environmental requirements for water.
>
> We face a governance crisis, rather than a water crisis. Water governance is about putting Integrated Water Resources Management, IWRM, with river and lake basin management and public participation and community empowerment as critically important elements of it, into practice.

How then is governance of river basin management improved? It involves using dialogue processes within a number of governance tools. These can be grouped into the following categories:

- Planning tools: river basin management plans, sub-basin and local land and water management plans (LWMP, as discussed in Chapter 4);
- Information capacity building tools: modelling, information exchange, geographic information systems (as discussed in Chapters 5 and 6);
- Social capacity building tools: leadership, advocacy, training, awareness and responsibility, multi-disciplinary team building, brokering agreements (Section 7.4);
- Institutional capacity building tools: the role of institutions, economic incentives, cost-sharing arrangements, pollution trading permits, covenants, rules and regulations and others (Section 7.5).

Section 7.3 is a discussion about what constitutes 'good' governance of river basins and the social and institutional capacity building tools which can assist governance.

First, let us look at several 'snapshot' examples of river basin commons and how some of these tools have been or can be applied in two contexts: developing country and developed country settings. These are short vignettes of typical river basin management problems and scenarios in two contrasting environments, those of an emerging economy of the South—the State of Andhra Pradesh, India—and various highly developed economies of the North.

7.2 BASIN GOVERNANCE SNAPSHOTS: SOUTH AND NORTH

Snapshot 1. Contrasting urban periphery environments in two river basins.

A rapidly urbanising urban periphery 'common': outskirts of Hyderabad, Andhra Pradesh—Krishna River Basin, India.

Rapidly changing landscapes on the rural–urban fringe of Hyderabad include newly constructed high-rise apartments and research parks, random squatter settlements which grow by encroachment and improved single carriageway road systems. Small pockets of intensive padi and tree crops remain, but much of the original forest vegetation was cleared during the last three decades. The monsoonal environment of this region receives the majority of its rainfall in the wet monsoon, much of which percolates into the highly porous granitic landscape. The photo shows recreation areas and ponds being constructed to harvest monsoonal rainwater, which in turn feed aquifers to sustain the growing urban populations (groundwater wells). The response by governments, where land-use planning is negligible, is to encourage wherever possible by voluntary means, management methods which ensure underground water resources are kept unpolluted and where possible, to fine negligent polluters. This is difficult in such areas where population growth is very fast.

A rapidly urbanising urban periphery 'common': outskirts of Madrid—Tagos Tajo Basin, Spain.

Rapidly changing landscapes (newly constructed apartment complexes, golf courses, freeways) reshape the hydrology of this watershed on the outskirts of Madrid. Which institutions can manage this rapid land-use change, to minimise impacts on the quality of the natural resources base of the river basin? Which governance protocols can be used? Land-use planning laws provide an opportunity to reduce peri-urban non-point source pollution. The role of the state and regional planner is important: to ensure regional growth planning mechanism are linked to river basin management processes which will ensure that development is done in ways which minimise environmental impacts. Local government planners can use such tools as land-use regulation, land taxes, land assessment and evaluation, best management practices for stormwater management and non-point source pollution management to control urban growth. Optimal effluent and stormwater quantity standards can be attached to a subwatershed requiring developers to reach environmental performance targets.

Snapshot 2. Contrasting metropolitan city 'commons' in two river basins.

A rapidly urbanising Asian city—Hyderabad, Andhra Pradesh—Krishna River Basin, India.

A rapidly urbanising western city—Chicago—Great Lakes Basin, USA.

The urban landscape of this city is one characterised by rapid change. The metropolitan population of Hyderabad is now over 5 million people, and continues to grow at an alarming rate as a result of rural–urban drift. The city has experienced a hi-tech industrial boom and service industries are rapidly overtaking industrial activities. The sheer size of the population, the lack of maintenance of an ageing wastewater infrastructure, problems with lines of accountability to capture and fine polluters, and the emergence of a widespread informal water sector which provides both tanker and bottled water, make the management of both water quality and quantity major problems. The city is looking elsewhere beyond the traditional tank system established to supply a much smaller population. Like many cities of the South, too much rainfall in one season and too less in another, the lack of capacity to store and mange water makes basin management a major challenge.

Water quality rather than water quantity dominates the agenda of natural resources management in the Chicago metropolitan region. This contrasts to those problems of cities of the South. The water sector is highly developed and a range of institutional arrangements such as penalties for pollution, service provision charges for water supply, a highly developed metropolitan water delivery system and rapid response mechanisms to address breakdowns. Cities like this of the North, particularly in temperate regions, are blessed with infrastructure which can be renewed (there is the financial capacity through national and state grants) and there is an abundance of precipitation which is more evenly distributed through the year. However, Chicago like other Great Lakes Basin cities is concerned about growing demands for freshwater extraction which will lower lake levels and create lakeshore management problems.

Snapshot 3. Contrasting industrial commons in two river basins.

An industrial commons in Jeedimetla, Andhra Pradesh—part of the Krishna River Basin, India.

A post-industrial commons—the Sydney Olympic Park, part of the Sydney Harbour Catchment, Australia.

The industrialisation of Andhra Pradesh is widespread but differs in character to that of countries of the North. Here, there is a preponderance of small-scale operators and economic conditions are marginal at best, nor is there the opportunity to use regulatory measures to enforce pollution. The use of a polluter pays approach is constrained by lack of enforcement and it would not only put operators out of business, but would have a roll-on effect and create local poverty; so there is reluctance to use such a process. The photo shows part of the solution—an externally funded wastewater treatment plant (by AusAID, the Australian aid agency) which provides a central wastewater treatment service for surrounding industries (leather tanning, light manufacturing). Wastewater is trucked to the site for treatment. These measures reduce serious pollutant impacts on groundwater systems (used for drinking and personal hygiene) and river systems.

Many former industrial areas in countries of the North have become derelict wastelands or areas of urban reinvigoration. In these post-industrial societies, the focus of IRBM is restoration, clean-up and creation of new landscapes ('building nature' as it is called on the Lower Rhine in The Netherlands). The above photos demonstrate this change dramatically in a harbour/estuarine environment. With significant capital investment for the Sydney 2000 Olympics, a large-scale derelict industrial region (left) of inner Sydney was transformed into a global sporting and recreation landscape (right) which included estuarine rehabilitation and containment of toxic land waste. Later, the Sydney Catchment Authority specified interim water quality and river flow objectives for the Sydney Harbour and Parramatta River catchment. The water quality objectives are for protection of aquatic ecosystems, visual amenity, and primary and secondary contact recreation. Using improved stormwater management, water quality will be maintained to achieve safe boating and swimming—particularly in areas of the Harbour to be used for Olympic events. The area is also managed for fishing after many years of severe pollution which restricted use. The Sydney Harbour intertidal zone is a declared Intertidal Protected Area under the Fisheries Management Act 1994, and intertidal invertebrates are protected from harvesting.

Sources: www.epa.nsw.gov.au, www.sydneyolympicpark.com.au

Snapshot 4. Contrasting agricultural commons in two river basins.

A highly populated, monsoonal agricultural 'common' in Warangal, Andhra Pradesh—Krishna and Godavari River Basins, India.

A depopulating salinised agricultural catchment—the Tragowel Plains of Northern Victoria, Murray–Darling Basin, Australia.

How is IRBM implemented in agricultural regions of countries of the South? They are characterised by rapidly growing populations, the demand for increasing production of food and fibre products to sustain local populations and increase exports to nearby regions of for export. The governance of IRBM must be done in ways which recognise the very large number of small-scale operators and few large-scale family farms or corporate farms, as are found in agricultural regions of the North. This requires a system of local governance driven by water user associations and other district level water initiatives, as well as poverty reduction, health and literacy programmes. The challenge is to link these processes with higher levels of governance in water, including large-scale water projects which can overrun local interests and lead to displacement. Local initiatives can be used to specify best management options which are affordable, employ large numbers of farm labourers and which provide food and income security.

How is IRBM implemented in agricultural regions of countries of the North? They are characterised by large-scale corporate farms and smaller-scale family farms. Populations have dramatically reduced from their peak in the 20th century. Production costs have risen and in many agricultural regions where subsidies do not exist, there has been widespread exiting from agriculture. The challenge in many agricultural regions is to implement sustainable farming systems, by implementing regionally determined best management options which require individual farmer effort coupled to basin-wide programmes of support. In this photo, salinity management plans have been introduced which focus on a structural adjustment programme to exit farmers from the system, restore degraded farming lands (using salinity hazard mapping, tile drainage, fencing off highly salinised lands, improved irrigation practices and removing degraded lands from production).

7.3 'BEST' PRACTICE IN IRBM GOVERNANCE

The snapshots show something of the diversity of river basin management problems and solutions. There is no one solution which will suit all, rather a suite of solutions need to be used and applied selectively to match the specific situation. In short, there is no 'holy grail' of IRBM process, no governance rules which suit all settings. Nor is IRBM a panacea for all basin scale natural resources management problems. In fact, in some basins, cultural diversity and political problems made preclude any advance. Stable democratic processes and institutions need to be established before IRBM can be developed. It is impossible to establish international and national IRBM if war and conflict exist. Correira and da Silva (1999), for example, cite nine international river basins undergoing conflict: Beas-Sutlej and Ravi (India and Pakistan); Brahmaputra and Ganges (Bangladesh, Bhutan, India and Nepal); Colorado and Rio Grande (Mexico and USA); Euphrates-Tigris (Iraq, Syria and Turkey); Lauca (Bolivia and Chile); Niger (Guinea, Niger, Nigeria, Mali); Nile (Egypt, Ethiopia and Sudan); Paraná and Rio de la Plata (Argentina, Brazil and Paraguay); Yarmak (Jordan and Syria), and maintain that international agreements are still vague legally, especially concerning allocation of water resources for different uses. Without definition of the agreements and resolution of conflicts over water use, little progress will be made in IRBM.

But there are lessons that have been learnt which drive 'best practice'. The attributes shown in Table 7.1 were designed for settings in both or either developing and/or developed country scenarios and are presented as an embracing list of 'best' practices.

River basin organisations and river basin scale natural resources management initiatives (where no RBO exists), as listed in Chapter 2, evolve with time and their composition, roles and responsibilities change over time reflecting the needs of the basin. This is an adaptive approach, as discussed in Chapter 3, and is appropriate to all settings.

The first step to establish good basin governance is to create an institutional environment which enables the creation of a river basin organisation relevant to its context. The second and subsequent steps are to establish an institutional framework and to apply management tools to enact IRBM. This ongoing process of institutional strengthening should be seen as a learning process, one driven by dialogue, where lessons are learnt step by step from real experiences. These attributes cover a broad spectrum of activities and it is unrealistic to assume all will be or need to be applied at once. It is more likely that considerable time will need to be spent creating an enabling environment in this first instance which can facilitate further basin governance measures.

The approach used here is to group attributes of best practice into three components:

- *The Enabling Environment* which comprises the laws, investments and policies of the nation or nations where the basin is located and which form the framework for natural resources management and development;

Table 7.1 Attributes of good governance in IRBM

Attribute	Description
1. Enabling environment—Laws, investments and policies	
International reciprocity and goodwill	For international basins, there will be international agreements which precede the establishment of joint basin organisations; stability of these international relationships will be the hallmark of preconditions to establish cooperative governance
Stable, democratic conventions	There exists a relatively stable set of institutions of government which provide the ability of the public sector to establish a system of policies, laws and financing arrangements, which are unhindered by civil unrest and democratic election of officials progresses between administrations without undue calamity
Functional specificity	Fragmentation and overlap of responsibilities is addressed by supportive legislation, clear specification of roles and responsibilities of basin partners
Clear management roles and jurisdiction	The policy and legislative framework will govern the purpose and effectiveness of the RBO, but an RBO requires a clear management role and jurisdiction
Problem scope specificity	Definition of the scope of the problem-shed, range of issues, environmental policies and management activities occurs; a clear boundary of the edge of the problem domain is established
Financial and human resource availability	The river basin management process and its RBO (if it exists) will have available: —adequate financial resources adequate to make substantial decisions which address priority natural resources management issues —well-trained staff with capacity to work in teams and plan across sectors and disciplines will be employed to implement these arrangements/RBOs
Private sector involvement	There is ample opportunity for the private sector to enact river basin management functions, especially at the local level. This can be realised through joint ventures, cost-sharing arrangements and common projects
Goal shift	Effective integrated river basin management moves from a pure resource exploitation ethic to incorporate social equity and environmental management in its work plans

(*Continued*)

Table 7.1 Attributes of good governance in IRBM (*Continued*)

Attribute	Description
Accountability	The enabling environment is supported by strong and comprehensive, but flexible legislation, regulations, decrees, etc. which ensures 'fairness' in basin-wide decisions and a process of accountability. 'Policing' by an independent body (or bodies) with enough authority to insist on improvements
Legal and jurisdictional setting	Need for established and accepted basin rules or laws including the legislation which clearly identifies its functions, structure and financial base; regulatory mechanisms and enforcement processes; trained staff who know and can use the law
Legislative back-up	Have a strong foundation and mandate in legislation which is based on a decision-making process, characterised by authority, responsibility and accountability
Realistic goal formulation	There is the need for: —well-defined objectives for river basin management with mutually beneficial and desirable goals, and where resource development forms part of a long-term IRBM plan —awareness of constraints on development in basin; awareness of 'turf' disputes —a strategic planning and implementation process are needed based on communications, coordination and cooperation —realistic and informed understanding of what are the feasible options
Failure to establish the need, scope and context	Specification of what are the priority natural resources management issues, how they can be addressed and a thorough understanding of the basin's hydrology will be part of the goal formulation

Table 7.1 Attributes of good governance in IRBM (*Continued*)

Attribute	Description
2. Institutions and capacity building	
Cross-sectoral integration	Integrated action used across all natural resource issues, which means agencies do not find singular solutions but look at impacts and improvements across the spectrum of natural resources, and the development of regional (basin scale) natural resources management policies
Coordination	Avoid bias in monitoring, planning and management through coordination of a range of state, national, international, commercial and private NGO bodies; use of a multiple agency approach with overarching coordination function by the RBO
Organisational style	Satisfactory organisational structures which allow cross-sectoral planning and management; focus on coordination and advisory roles; the basin entity focuses on oversight, management and planning
'Balanced', agreement-driven management	Basin-wide planning is used to balance all user needs for water resources and to provide protection from water-related hazards; agreement on commitments within the basin, and mechanisms for monitoring those agreements
Local and regional planning capacity	The RBO guides and coordinates local government agencies to enact zoning mechanisms, local government pollution controls, planning tools to manage local natural resources, congruent with overall basin management goals
Rules governing structure	The position and boundary of the RBO's rules are clearly defined; specification of the entities involved in IRBM and their roles in decision settings; 'rules' of participation specify roles of participation and membership and exiting decision settings
Clear coordination process	Rules are defined for the array of coordination activities (who is involved), how binding or permissive is the coordination (what can be done) and on what basis is the involvement (law, policy, informal agreement)
Rules governing process: decision rules	Specification of how decisions are made (consensus, voting, etc.)

Table 7.1 Attributes of good governance in IRBM (*Continued*)

Attribute	Description
3. Management instruments	
RBO leadership	RBOs have the mandate to ensure they tale the 'big picture' in river basin management; they provide the leading voice on basin-wide water issues; they inform their constituencies and decision-makers in all sectors and at all levels of decision-making in both the public and private sector; decision-makers will be well-trained, articulate, responsible and listening skills
Realistic management	RBOs make decisions aware of the reality of existing conditions; often compromise on the best management options is required; a staged implementation procedure is needed—addressing the most pressing resource management issues first, and recognising what is possible in the short-term; this process must be backed up by long-term planning
Public sector leadership	Water, as a common-pool natural resource, will more likely be managed by the public sector, than the private sector; the state (both provincial and local agencies) will take the lead role to develop, implement and manage river basin management activities
Knowledge system	RBOs require a high-quality, reliable, uniform and comprehensive data network, available to all stakeholders in ways which suit their needs; systems and models for analysis which allow 'knowledgeable' natural resources/water management policies and strategies to be developed and implemented
Jurisdiction over an informal water sector	It is meaningless to consider regulatory instruments over water use when there are vast numbers of small-scale users and ground water pumpers who are not linked with public institutions; 'rules' (agreed uses) rather than regulations are more likely to succeed
Improving the productivity of 'Green Water'	Increasing the productivity of water diverted from rivers is less important than being able to capture water more effectively in the soil profile; mechanisms for raising local productivity through village-led local initiatives in water harvesting are the fundamental tool for local water management

Table 7.1 Attributes of good governance in IRBM (*Continued*)

Attribute	Description
Address water scarcity using a cross-sectoral approach	Very large and rapidly growing populations depend on a limited natural resource; the challenge is to get more crop, cash and jobs for each drop; basin-wide water management is linked to securing safe, clean and accessible water supplies and linked to family planning and health programmes
Information accessibility and integration	Information is provided in an integrated, interpreted form; resource managers do not necessarily need raw data, but information and knowledge about what works best and where—informed by the latest science, resource engineering, resource economics and practical experience
Water pricing and demand management	Pricing mechanisms are best applied to contexts where mechanisms for water charges can be collected; the price of water retains a poverty clause to provide water as a fundamental human right; alternative demand management technologies are used where pricing is in appropriate and used in conjunction with pricing where users have a capacity to pay
Rules governing information exchange	The content of the information used by participants is specified; the form of the information and the timing of information exchange is known; the methods of exchange are accessible, appropriate, equitable and affordable
Research system (data and monitoring)	A well-designed research programme which informs all stakeholders of best management options for land types in sub-basin catchments; the programme is aided by the provision of data, monitoring and understanding of the basin structure and function and resource activities
Effective community participation	Strong community awareness and participation processes exist to enhance greater ownership of basin scale plans of action; the emphasis is placed on wide public and stakeholder participation in decision-making at all levels; local empowerment is facilitated if participation is a high priority
Flexible, adaptable management	Use a 'learn by doing' and 'development facilitator' approach—makes planning and management more adaptable

Source: Adapted from Barrow (1998), Bellamy et al. (1999), Hooper, McDonald and Mitchell (1999), Millington (1999), Alaerts (2001), Global Water Partnership (2002), Global Water Partnership (2003), Van Steenbergen and Shah (2002), Shah, Molden and Sakthivadivel (2003a).

- *The Institutional Framework* which specifies the institutional arrangements and organisations to manage IRBM; and building the capacity of the institutions and organisations to deliver arrangements and practices for IRBM;
- *The Management Tools* which comprise a number of specific tools for improved governance of river basins such as community participation, water resources assessment and modelling, river basin plans and local government plans, education, communication and awareness raising, economic instruments, regulatory tools and information exchange.

These practices were drawn from the extensive, international experience of practitioners and researchers in river basin management who are listed below Table 7.1. The approach is derived from the theoretical construct of governance developed for IWRM (Global Water Partnership Technical Advisory Committee 2000; Global Water Partnership 2003; Rogers and Hall 2003) and which was used in the IWRM ToolBox of the Global Water Partnership (Global Water Partnership 2002). As discussed in Chapter 1, IRBM is seen as a subset of integrated water resources management.

7.4 SOCIAL CAPACITY BUILDING TOOLS FOR IRBM

Social capacity building has been shown to be a powerful tool to implement IRBM. Leach and Pelkey (2001), for example, analysed the factors affecting conflict resolution in watershed partnerships in the United States. In 37 available studies, they identified 210 'lessons learned', which were grouped into 28 thematic categories. The most frequently recurring themes were the necessity of adequate funding (62% of the studies), effective leadership and management (59%), interpersonal trust (43%) and committed participants (43%). Using factor analysis, they found factors such as balancing the partnership's resources with its scope of activities and employing a flexible and informal partnership structure were critical.

In a much broader application (Anonymous 2000), a collaboration of NGOs called for recognition of the marginalisation of society's socially weaker groups, rural dwellers, indigenous peoples and the urban poor by river basin management and poor water resources management. Their vision for water built on the insights and views which surfaced during regional consultations undertaken in Kenya, India, Brazil and Poland. Their message, while addressing problems and opportunities which prevail in the South, bears a message for river basin and water management in the northern hemisphere: that social concerns are fundamental to basin management.

Both studies and several others (Allee, Apener and Andrews 1975; Wolman 1981; Saeijs and Turkstra 1994; Shuval 1994; Cameron 1997; Cantwell and Day 1998; Harmsworth 1998; Cosgrove and Rijsberman 1999; Letey 1999; Mullen and Allison 1999; Lamoree and Nilsson 2000; Barham 2001; Bellamy et al. 2001; McCool and Guthrie 2001; Oliver 2001; Parkes and Panelli 2001; Walker 2001a;

Walker 2001b; Falkenmark et al. 2004) attest to the fundamental need to enhance IRBM using social capacity-building tools—these are not 'add-ons' to river basin projects, but the 'core business' of IRBM. These practitioners and researchers call for lasting social institutions which transcend political and administrative boundaries and which generate direct benefits to people from IRBM; they plea for stronger recognition of the social domain in river basin management and advocate a stronger participation by IRBM stakeholders at the highest levels of decision-making.

7.4.1 Leadership skills

Leadership is a critical factor ensuring the success of IRBM. Visionary leadership can be the key to ensuring clear IRBM products. Without this vision it is not possible to have a meaningful planning process. When there is community leadership and a shared vision across a range of sectors, it is more likely that river basin management process will succeed. Leadership can be difficult and many strong egos potentially lead to conflict if focus or momentum is not maintained.

Strong RBO leadership is valuable as it brings together community, private sector and government interests by working in a partnership manner with common objectives. This will result in a high degree of trust in the RBO, especially if it is lead by capable chairmanship—people with political acumen and who can engage both willing and combatant natural resource managers. However, dependence on capable leaders can be problematic. Dependence on one individual with strong and capable leadership skills can lead to river basin management projects being vulnerable without a strategy for leadership succession.

In many situations, RBO leadership should be coupled with strong leadership in the local community, and thus linked to the institutional arrangements for IRBM and a knowledge base of 'good science'. Local communities include indigenous groups. In Aboriginal communities in the Northern Territory of Australia, success in integrated resource management programmes occurred because of the integrating skills of Aboriginal facilitators and community leaders, to form an holistic approach to resource management and community development. The Julalikari Council Aboriginal Corporation of Tennant Creek adopted this role and called themselves 'programme brokers' (AACM International and Centre for Water Policy Research 1995a).

What is critical to IRBM success is:

- strong leaders with an authoritative, visionary and embracing leadership style;
- leaders with conflict resolution and time management skills;
- information systems to provide leaders with unambiguous information about best practice, data on the financial status of their organisation and access to key political, industry and community people;
- leadership training facilities;

- a processes for initiation and early development of community group structures, to allow groups to mature into independent, self-determining institutions;
- financial incentives (at CEO levels) to engage leaders to remain involved in river basin management.

(AACM International and Centre for Water Policy Research 1995b)

Where river basin management depends on voluntarism, there is a tendency for burn out by leaders, so there is the need for a process of leadership succession and the need to invest public resources and support mechanisms to provide programmes to assist leadership development.

7.4.2 Political advocacy

Political advocacy is defined as the range of activities by action groups which seek to influence political decisions, for their own ends. Citizen, media and NGO pressure frequently galvanises action: for example, to reduce major international basin management problems from water overuse (e.g. Aral Sea). The activities of such groups include influencing high-level government policy development, lobbying governments to pass legislation on river basin policy initiatives, to 'pork barrelling' by elected representatives to win favour in their electorates. There is a range of actions from the subtle to the more obvious. RBOs themselves can also function as river basin advocates on major national and international natural resource management issues.

These influences may generate conflict or, if congruent with those of other interest groups and the broader general public interest, resolve river basin management issues and assist political processes. Pressure groups, NGOs and the media can influence the decisions of politicians, cabinet groups or other high-level steering groups and government department heads, and so cause governments to create policy initiatives and new legislation.

IRBM operates usually in a highly charged political environment because its concern is the vital resource, water. IRBM is best done when it has leadership which is closely connected to the processes of central governments. One way which is helpful is for RBOs to be the 'mouthpiece' for the basin. This cannot be done by relying on voluntary ad hoc approaches, but it is more likely to be effective if the RBO has a direct reporting mechanism to natural resources ministers or cabinet subcommittees for natural resources management. The RBO has clear lines of communication to strategic high levels of government and sets a regional focus for natural resources planning and economic development.

Figure 7.1, for example, illustrates the structure for the Murray–Darling Basin Commission in Australia, which reports to a separate ministerial council, made up of the natural resources ministers of the states of the basin and which also includes a Commonwealth (central) government ministerial representative. In this way, the Commissioner of the Murray–Darling Basin Commission has to account for the Commission's performance in fulfilling its annual objectives directly to the highest

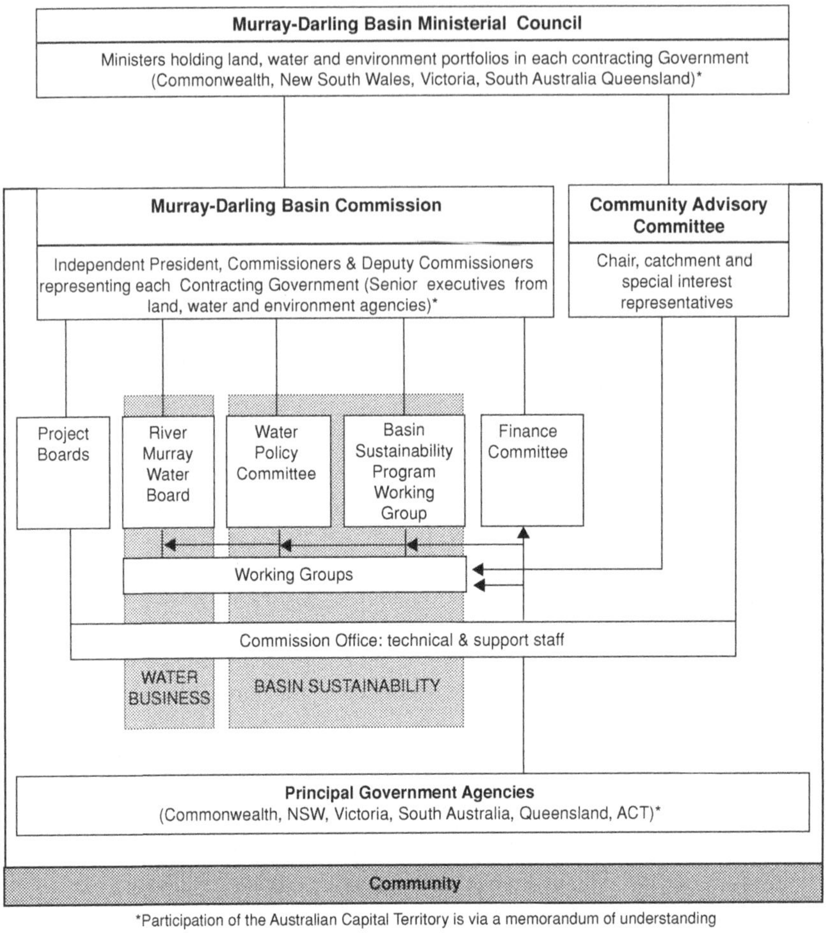

Figure 7.1 Governance of the Murray–Darling Basin Initiative.[1]
Source: Murray–Darling Basin Commission.

levels of government in the country, and can also advocate to government on the Commission's and the Basin's behalf.

The Ministerial Council's main functions, specified in the Murray–Darling Basin Agreement, are

- Generally to consider and determine major policy issues of common interest to the Contracting Governments concerning effective planning and management for the equitable efficient and sustainable use of the water, land and other environmental resources of the Murray–Darling Basin.

[1] The Murray–Darling Basin Initiative is the strategic work programme of the Murray Darling Basin Commission.

- To develop, consider and, where appropriate, to authorise measures for the equitable, efficient and sustainable use of such water, land and other environmental resources.

  Source: www.mdbc.gov.au, accessed December, 2004.

Similarly, the Community Advisory Council has a direct advocacy role. Its terms of reference are to advise the Ministerial Council on:

- the natural resource management issues referred to the Committee by the Ministerial Council
- the full range of views of Basin communities on natural resource management issues of significance within the Basin.

As well, it assists the Murray–Darling Basin Initiative by disseminating within Basin communities, Ministerial Council's decisions in a way that promotes clear understanding of their context and rationale, and enhances their ownership and adoption. It participates, as directed by Ministerial Council, in Basin community engagement programmes and provides the Ministerial Council with advice on the effectiveness of that engagement. It also participates in policy development processes of the Commission and Ministerial Council. (Source: www.mdbc.gov.au, accessed December, 2004.) The CAC is composed of 20 members plus the independently appointed Chairman. Members of the CAC are appointed for 4-year terms and are selected on the basis of their skills and expertise (in five key sectors—irrigation industry, urban, dryland farming, environment, local government) and networks (indigenous members and representatives of each of the member states).

This example serves as a practical, valuable model of basin advocacy for other RBOs.

7.4.3 Training a new generation of river basin managers

In Chapter 1 of this book, I described IRBM as a new paradigm in river basin management. The methods of river basin governance outlined throughout this book will be 'new material' for many natural resources management professionals. For others, hopefully, this book may echo their experiences and point them in new directions with new insights. IRBM is not a simple method of natural resources management and requires expertise from many areas. Therefore, it cannot be simply distilled into a neat programme of instruction applicable to all contexts.

Training river basins mangers is a critical task as it is in the next generation of managers that change will occur. Broadly speaking, river basin mangers need training in natural resources management, resource economics, resource engineering, social sciences and human resources management (especially multi-disciplinary team building and dialoguing skills). This comprehensive curriculum will not be achieved in one tertiary level course (say a Masters degree in integrated river basin management), nor a short course offered to basin professionals or a specially designed course tailored to a specific setting. Rather, a combination of these through a person's professional career is what is needed. One specification of IRBM education suggested earlier is to:

- Provide specific courses on participatory approaches and gender awareness.
- Encouraging multi-disciplinary training involving all kinds of water practitioners, including environmentalists, economists, engineers, social scientists and business leaders.
- Including water management in degree programmes, in engineering and other faculties, such as economics, environmental sciences, biology, etc., or adding water as a major in such degree courses as an MBA.
- Developing modules for on-the-job training to keep practitioners' skills up-to-date.
- Developing training of trainers modules in new approaches and techniques.
- Creation of short courses on water management for policy-makers, aimed specifically at senior managers without technical water backgrounds.
- Once formal training is completed, the concepts can be reinforced through a range of training activities (e.g. on-the-job training, short courses, remote learning, sabbaticals, twinning arrangements, international short courses, etc.).

  Source: Global Water Partnership (2002), Tool: B2.2.

One of the powerful techniques required of river basin managers today is the ability to dialogue, communicate and effectively share information, knowledge and wisdom. There is the need to train river basin management professionals in information exchange and communication using methods such as in-service courses, seminars and workshops (Lewicki 2001). As many information exchange facilitators (such as extension officers, field guides and field agents) come from biophysical science and engineering backgrounds, one of the big challenges in preparing the next generation of river basin managers is cross-disciplinary training in communication, group interaction facilitation, cost accounting and programme management.

The Global Water Partnership notes the following lessons learned in training water professionals in integrated water resources management. These apply equally to training river basin management professionals:

- Training of senior managers (e.g. in the value of IWRM and new water innovations) can help ensure capacity building throughout the organisation, and support for training of junior staff.
- On-the-job training is highly effective as a learning tool and agent of change in large water organisations.
- The effectiveness of training programmes can be increased if groups of people that regularly work together are trained together.
- Training of trainers requires extensive practical experience by the instructor but is a cost-effective capacity building tools.
- Trainers do not require a high level of technical capability in such topics as how to construct GIS, develop explanatory models or select the best equipment, but

they do need to understand the management of institutional and organisational structures.
- Experience shows that successful courses to train trainers combine learning by doing with classroom learning experiences.
- Regional and even international programmes can be as useful as programmes that focus only on a single country or state.

Source: Global Water Partnership (2002), Tool: B2.2.

7.4.4 Awareness and responsibility, multi-disciplinary team building and brokering agreements

The task of IRBM is essentially one of the coordination across different sectors in natural resources development and management. The critical role of the river basin manager is to foster a spirit of coordination amongst stakeholders, then move to establishing methods for co-governance. This involves two management tasks.

Building awareness and responsibility

In this task, the RBO plays a lead role in raising awareness of critical basin problems and the need for stakeholders to take hold of their responsibility in solving these problems. This task can be a major stumbling block to coordinated management as stakeholders:

- may not see themselves as being located in the basin, or they are unaware such a basin exists (see Chapter 2)
- may be ignorant of the idea of a river basin, what is natural resources management, and what is IRBM—just another 'buzz word' of government and the media?
- may consider their role to be miniscule and irrelevant, compared to the large scale of the basin, so their individual contribution to solving problems is not worth the effort of involvement.

The development of individual and corporate responsibility amongst river basin stakeholders is predicated on a well-orchestrated awareness raising campaign. Slogans such as 'we all live in a river basin' help to raise this basin-wide awareness and should be considered as a first step towards ownership of problems. Public access to information is also a powerful mechanisms in raising basin awareness. Here, there is a critical role for the RBO to play—to have an open, transparent system of information availability for basin stakeholders. This can be done by many mechanisms including:

- Direct use of conventional media (printed media, TV, radio) and/or non-conventional media (messages on water bills, games, transport tickets, comic books, etc.)
- Organisation of large events and the endorsement of celebrities (generating media attention)
- Use of existing networks (religious networks, social movements, NGO networks, business associations)

- Stencilling rivers and creeks within a basin on road signs to show they are part of a larger basin; marking river basin boundaries on highways, similar to state and international boundaries
- Use of logos (e.g. a water drop) to give identity to the campaign.

Source: Global Water Partnership Technical Advisory Committee (2000), Tool: C4.3.

These tasks are broad scale in impact and the results are often only of marginal importance to the most important stakeholders. An RBO has the opportunity to be the lead agent with these stakeholders by personal invitations to meetings and building strong personal relationships with CEOs, community leaders, water mentors and gurus and other change agents in business and the community in the basin. This will open the door for awareness raising but also inculcating an ethic of ownership and responsibility for river basin management. This suggests that RBOs must have staff trained in the skills of public relations, information exchange and dialogue.

Multi-disciplinary team building for national basin planning

Allied to and resulting from the sense of ownership which comes from knowing a river basin's natural resources problems, there is the opportunity for RBOs and basin initiatives to build multi-disciplinary teams of resource managers and stakeholders to cooperatively plan and work together to solve problems.

The Wisconsin Department of Natural Resources used teams of resource managers and stakeholders with differing and complementary specialisations to solve natural resources management problems on a watershed basis. Margerum (1995) noted the teams used a number of coordination techniques for implementation including:

- Regular meetings of people involved in implementing a plan
- Regular conference calls or memoranda
- The team met regularly and was called together to when specific issues or problems arose
- Shared databases or information systems.

He recommended from this experience that:

- The planner/coordinator should have facilitation and conflict resolution training.
- The collaborating team should develop clear operating procedures.
- Collaborating teams should decide how it will resolve conflicts.
- The planning phase should be well documented.
- Stakeholder participation should proceed parallel to public input.

AACM International and Centre for Water Policy Research (1995b) in a national consultancy project to improve the performance of Australian catchment management noted examples of multi-disciplinary teams which:

- solved catchment management problems, while promoting the integrated catchment management concept, and culture change in natural resources management organisations which emphasised coordinated action
- linked state and local government staff with community groups to design, implement and monitor catchment management activities
- harnessed pre-existing links between agencies to collect and share data
- established links between traditional owners, technical services agencies and catchment management councils to develop and implement planning processes and develop best management practice guidelines for different land uses.

These experiences led the AACM consultants to recommend to the Australian government that multi-disciplinary teams should be integral to a national mechanism to strengthen integrated catchment management. They noted that,

> Multi-disciplinary teams provide a means of integrating different skills, and establishing working relationships and communication between different government agencies. This approach integrates institutions horizontally and vertically. Whilst in many regions of Australia this approach is used in an informal way, it is rarely adopted as a formal component of the integrated catchment management process.
>
> (p. 40).

7.5 BUILDING INSTITUTIONAL CAPACITY

7.5.1 What are institutions?

There are various institutional arrangements to enact IRBM. The following discussion points to a sample of IRBM institutional arrangements, believed to be critical to developing effective IRBM. Basin managers need to explore institutional arrangements which fit their context and an initial scoping of the enabling environment (policies, laws and financing mechanisms) should always precede the development of more specific institutional arrangements.

> **Institution**—the existence of formal or informal rules that govern the actions of that institution.
>
> *Source:* Scott (1995).

7.5.2 Brokering strategic, high-level agreements on river basin management

The ability to enact IRBM requires decisions being made at the highest levels of governments and between governments for international basins. For the latter, these are often preceded by international agreements about water sharing and international cooperation on other issues, such as trade and economic development. The challenge in both national and international settings is to work towards agreed programmes of action rather than actions which cause separate uncoordinated decision-making and which frequently result in conflict. The implementation of

IRBM requires the provision of national land and water resources management policies and strategic planning techniques, and investment strategies.

International settings

With respect to international river basins (as listed in Annex 1), the United Nations Convention on the 'Law of the Non-navigational Uses of International Water Courses', adopted in 1997, is an excellent starting point for negotiations on water resource sharing and management on a river basin basis. It is based on the principles of 'equitable and reasonable utilisation' of water resources.

Article 5 for example states that,

1. Watercourse States shall in their respective territories utilize an international watercourse in an equitable and reasonable manner. In particular, an international watercourse shall be used and developed by watercourse States with a view to attaining optimal and sustainable utilization thereof and benefits therefrom, taking into account the interests of the watercourse States concerned, consistent with adequate protection of the watercourse.
2. Watercourse States shall participate in the use, development and protection of an international watercourse in an equitable and reasonable manner. Such participation includes both the right to utilize the watercourse and the duty to cooperate in the protection and development thereof, as provided in the present Convention.

Article 6 goes on to list the factors relevant to equitable and reasonable utilisation,

1. Utilization of an international watercourse in an equitable and reasonable manner within the meaning of article 5 requires taking into account all relevant factors and circumstances, including:

   Geographic, hydrographic, hydrological, climatic, ecological and other factors of a natural character;

   The social and economic needs of the watercourse States concerned;

   The population dependent on the watercourse in each watercourse State;

   The effects of the use or uses of the watercourses in one watercourse State on other watercourse States;

   Existing and potential uses of the watercourse;

   Conservation, protection, development and economy of use of the water resources of the watercourse and the costs of measures taken to that effect;

   The availability of alternatives, of comparable value, to a particular planned or existing use.
2. In the application of article 5 or paragraph 1 of this article, watercourse States concerned shall, when the need arises, enter into consultations in a spirit of cooperation.
3. The weight to be given to each factor is to be determined by its importance in comparison with that of other relevant factors. In determining what is a

reasonable and equitable use, all relevant factors are to be considered together and a conclusion reached on the basis of the whole.

Article 7 focuses on the obligation not to cause significant harm to shared water resources between countries in a river basin,

1. Watercourse States shall, in utilizing an international watercourse in their territories, take all appropriate measures to prevent the causing of significant harm to other watercourse States.
2. Where significant harm nevertheless is caused to another watercourse State, the States whose use causes such harm shall, in the absence of agreement to such use, take all appropriate measures, having due regard for the provisions of articles 5 and 6, in consultation with the affected State, to eliminate or mitigate such harm and, where appropriate, to discuss the question of compensation.

Article 8 then expounds the mechanisms by which joint sharing of water resources can occur as a matter of 'general obligation' to cooperate,

1. Watercourse States shall cooperate on the basis of sovereign equality, territorial integrity, mutual benefit and good faith in order to attain optimal utilisation and adequate protection of an international watercourse.

   In determining the manner of such cooperation, watercourse States may consider the establishment of joint mechanisms or commissions, as deemed necessary by them, to facilitate cooperation on relevant measures and procedures in the light of experience gained through cooperation in existing joint mechanisms and commissions in various regions.

The basis for which this mutual cooperation can occur is built around regular exchange of information as stated in Article 9,

1. Pursuant to article 8, watercourse States shall on a regular basis exchange readily available data and information on the condition of the watercourse, in particular that of a hydrological, meteorological, hydrogeological and ecological nature and related to the water quality as well as related forecasts.
2. If a watercourse State is requested by another watercourse State to provide data or information that is not readily available, it shall employ its best efforts to comply with the request but may condition its compliance upon payment by the requesting State of the reasonable costs of collecting and, where appropriate, processing such data or information.
3. Watercourse States shall employ their best efforts to collect and, where appropriate, to process data and information in a manner which facilitates its utilization by the other watercourse States to which it is communicated.

Finally, special clauses are recognised in Article 10 regarding the relationship between different kinds of uses,

1. In the absence of agreement or custom to the contrary, no use of an international watercourse enjoys inherent priority over other uses.

2. In the event of a conflict between uses of an international watercourse, it shall be resolved with reference to articles 5 to 7, with special regard being given to the requirements of vital human needs.

   Source: http://www.un.org/law/ilc/guide/8_3.htm, accessed December, 2004.

This approach requires mechanisms to be established to share water resources between nations in international basins. Sharing is determined legally on the basis of 'equitable' and 'reasonable' shares. This suggests the need for river basin organisations to be created and operate under this and their own national laws to test these definitions.

However, the strategic, high-level nature of these international basin arrangements are not just driven by law. They are also supported by an open and functioning information 'trade' between member states to enhance IRBM. Hence, the emphasis in this book for information exchange (Chapter 5) and a prototype RBIS (Chapter 6) which could be established between international river basins member states under this jurisdiction. Similarly, the emphasis on decision-making (Chapter 3) underpins the need for international river basin management to be built on the design of good decision-making principles—those built on equitable and reasonable values.

National settings

There is also the need to establish guiding principles in national settings so that IRBM can be established in a similar fair manner. The principles of equitable and reasonable use suggest that in a national jurisdiction, member provinces and districts have the right to use water for beneficial use as well as a complimentary obligation to cooperate. This is done within national operating guidelines of water use—usually contained in national water policy. These principles can guide the development of an investment strategy for national governments whereby their member provinces can develop means to use and share water between and within them.

The Australian example shown in Table 7.2 is an integral component of institutional capacity building. AACM International and Centre for Water Policy Research (1995b) recommended the establishment of a high-level strategic investment strategy for natural resources management and that river basin and catchment management be funded by a mix of public and private funds. This approach can be applied to national river basins elsewhere, but the guiding principles would need modification to suit the level of economic and institutional development, especially financing options from donor aid agencies and other sources. It is important to recognise the role of multi-disciplinary teams as part of a national strategy for strengthening IRBM throughout a nation, and to strengthen coordination between stakeholders.

This approach has some similarities to that developed in the European Water Framework Directive for river basin management (Chave 2001), where river basin management plans are required and are paid for by cost-sharing arrangements

Table 7.2 Guiding principles for successful integrated catchment management

Clear investment framework A national Integrated Catchment Management Investment Strategy, based on resource economics, which clearly establishes priorities for Commonwealth and State investment in natural resource management as a framework for regional resource management planning activities
Cyclical resource management process A cyclical approach to planning which uses rolling renewal of programmes to allow dynamic responses to changing priorities and community perceptions whilst demonstrating a long-term commitment to integrated catchment management
Cost-sharing for co-management partnerships Clear co-financing of resource management activities on the land to establish a strong foundation for co-management partnerships between government and individuals. Use of resource economics to allocate public and private costs and benefits for different resource management activities
Contract for action Contracts—between incorporated community groups and landholders, technical services agencies, local government and public sector investment programmes—lead to open and sustainable co-financed management partnerships. Contracts would involve the development of appropriate cost-sharing, co-financing and co-management arrangements
Multi-disciplinary team approach Multi-disciplinary teams provide a means of integrating different skills, and establishing working relationships and communication between and within different government agencies. This approach integrates institutions horizontally and vertically
Strengthen with legislative frameworks A legislative framework is required to strengthen and formalise the process for coordination and management of resource management investments. It also provides a mechanism of last resort for minimising risks affecting outcomes expected from Commonwealth (national government) investments in integrated natural resources management

Source: Modified from AACM International and Centre for Water Policy Research (1995b).

and adopting a polluter pays approach. But there is no consistent international investment framework, except funds from the European Union to establish water resources management initiatives.

AACM International and Centre for Water Policy Research (1995b) went on to suggest that high-level brokering of strategic agreements for IRBM could be made:

> There are opportunities to change...... by integrating policy frameworks and institutional structures at the top (Commonwealth) or bottom (regional) levels. Integration in the middle (State) level without concurrent integration at top or bottom levels will not result in integrated resource management. Given the powerful incentive for change which national investment presents, the most efficient opportunity for change is likely to be integration of Commonwealth policy frameworks and institutional structures.
>
> There is an opportunity to develop cross-portfolio policy frameworks which integrate national resource management interests across Commonwealth portfolios and programs. These

actions should aim to eliminate contradictory messages to regional Australia, and the States, about Commonwealth priorities and national interests for investment in integrated resource management.

Opportunities exist for the leading national natural resources, primary industries and environmental agencies to develop joint initiatives and co-financing agreements for various components of an agreed national integrated resource investment strategy. This approach will also provide government with a mechanism, associated with the proposed contractual system, in which it can account for funding programmes believed to be in the national interest in natural resources management.

Similarly, at The Third World Water Forum in Kyoto, 2002, a gathering of water professionals and managers[2] from around the world agreed to the need for strategic planning in basin governance. They endorsed integrated water resources management and basin management as preferred approaches to meeting the world's water resources problems and recommended brokering high-level agreements and strategies for action. In Recommendation 3, participants agreed that,

> Multi-stakeholder partnerships at regional, country and local levels should be promoted, including the water related Type II Partnerships launched at WSSD in Johannesburg. Such partnerships include governments, private sector, academia, NGO's and civil society organisations.

In Recommendation 4, they called specifically for the creation and support of river and lake basin management strategies and organisation structures,

> As also stressed in the WSSD Plan of Implementation there is a need to "develop and implement national/regional strategies, plans and programmes with regard to integrated river basin, watershed and groundwater management", including "programmes for mitigating the effects of extreme water related events". Such plans should be flexible and dynamic and responsive to changes in society and climate.
>
> The creation and support to river and lake basin organizational structures involves all stakeholders, and include public participation through the mobilization and empowerment of the users and other relevant interest groups.
>
> The support required for basin management spans from policies and laws through regulations, standards, financial arrangements and information management to practical capacity building at all levels.
>
> (Anonymous 2002).

In short, this endorsement supports the need for basin capacity building through strategic plans in both international and national settings. It demonstrates the need to build IRBM by strengthening decision processes at high levels of governance.

7.5.3 A best practice paradigm for RBOs

A change in organisation performance and attitude is required by basin organisations to implement the emerging approach of equitable and reasonable sharing of waters by international basin members states and the development of basin-wide investment strategies within nations.

---

[2] The convenors of the IWRM session were the Global Water Partnership, International Network of Basin Organisations, International Council on Local Environmental Initiatives, Shiga Prefecture Government and the United Nations Environment Programme.

Table 7.3 The seven S's of organisational action[3]

The Hard S's	
Strategy	Actions a company plans in response to or anticipation of changes in its external environment
Structure	Basis for specialisation and coordination influenced primarily by strategy and by organisation size and diversity
Systems	Formal and informal procedures that support the strategy and structure (systems are more powerful than they are given credit)
The Soft S's	
Style/culture	The culture of the organisation, consisting of two components: Organisational Culture: the dominant values and beliefs, and norms, which develop over time and become relatively enduring features of organisational life Management Style: more a matter of what managers do than what they say. How do a company's managers spend their time? What are they focussing attention on? Symbolism—the creation and maintenance (or sometimes deconstruction) of meaning is a fundamental responsibility of managers
Staff	The people/human resource management—processes used to develop managers, socialisation processes, ways of shaping basic values of management cadre, ways of introducing young recruits to the company, ways of helping to manage the careers of employees
Skills	The distinctive competences—what the company does best, ways of expanding or shifting competences
Shared values/ superordinate goals	Guiding concepts, fundamental ideas around which a business is built—must be simple, usually stated at abstract level, have great meaning inside the organisation even though outsiders may not see or understand them

Source: http://www.themanager.org/Models/7S_Model.htm and Waterman, Peters and Phillips (1980).

Basin organisations work best when strategic alliances and agreements are formed. These can be assisted by new organisational style, based on mutual respect, empowerment, negotiated bargaining and consensus skills. This approach is similar to that recommended by the 'new environmentalism' emerging in the business community.

Any organisation can be visualised according to seven 'S' elements (Waterman, Peters and Phillips 1980) (Table 7.3). The 7-S model can be applied to demonstrate the organisational change required by basin organisations moving from a single sector to a coordinated, adaptive decision-making role to implement IRBM (Table 7.4).

[3] Waterman et al. distinguish between Hard S's (feasible, easily identified functions of organisations) and Soft S's (an organisation's practices which are difficult to describe as capabilities, values and elements of corporate culture develop and change through time). This division suggests any organisation is a dynamic, living organism, one which may be characterised by an outward structural formation but which is difficult to describe, let alone assess in terms of performance.

Table 7.4 The best practice environmental management paradigm shift for river basin organisations using IRBM

's' factor[4]	Old paradigm—single sector oriented natural resources management decision-making	New paradigm—a decision-making paradigm based on integrated, adaptive management
The Hard S's		
Strategy	Reactive —Meet regulations, focus on end-of-pipe —No specific environmental policy —Closed door to community	Proactive —Link between environmental excellence and competitiveness —Emphasis on continuous improvement —'Open door' to community
Structure	Rigid —Steeply hierarchical —Weak or no links between OH&S, environmental and production management	Flexible —Devolution of environmental responsibility —Flatter, team oriented —Integration of OH&S, environmental and production management
Systems	—Environmentally exclusive —Minimum required to meet regulations	—Environmentally inclusive —Comprehensive environmental management plan —Formalised communication links with community
The Soft S's		
Style	Formal —Command and control —Environment is a low priority of CEO	Committed —CEO vision, personal commitment and leadership —Demonstrated priority for senior management
Staff	Directed —Performance measured by cost —No sense of ownership —Pride in activities in IRBM	Empowered —Environmental criteria in performance appraisal
Skills	Functional —Production and waste control	Problem-solving —Integrated approach to improvement —Innovation, problem solving skills highly regarded
Shared values/ superordinate goals	Efficiency —Maximise business output at least cost with minimum expenses	Excellence —Strive for optimal river basin management outcomes, using stepped approach

Source: Adapted from Australian Manufacturing Council (1992).

[4] From Waterman, Peters and Phillips (1980).

7.5.4 Economic incentives

Financing river basin management is a perennial problem. As it deals with the commons of basin-wide natural resources, it is usually funded by public monies. However, there is growing recognition by governments of the need to recognise the role of private sector contributions in funding river basin management. This comes from the need to recognise that individual resource managers in many sectors derive benefits from public funded programmes and that public investment in natural resources management brings uneven benefits across basins. As a response, governments now recognise the valuable role of cost-sharing mechanisms (Section 7.5.5). There are a number of other financial and economic instruments available to IRBM. These come as part of broader natural resources and economic development government budget allocations: cost-benefit analysis (CBA) or more recent derivations where ecological economics and traditional CBA interact (Bruins, Heberling and Maddock 2005), multi-criteria analysis of basin management practices (including costing options), water pricing mechanisms, pricing of waste disposal charges and the use of taxes and penalties.

However, the core funding for river basin management organisations and water management has traditionally come from direct appropriations from a state's revenues (e.g. Dutch water management is paid by a mix of national government and provincial government appropriations, Water Board funds raised from the public fees, municipal taxes and semipublic water companies fees (Havekes et al. 2004); Federal Government appropriations pay works by the Tennessee Valley Authority in the United States). Foreign funding bodies such as the World Bank are the traditional funding sources for developing country basin organisations and basin management projects. Some basin organisations have used their rating powers (like the former catchment management trusts of Australia) to raise a basin levy for the funding of catchment works and programmes.

Economic analysis and social impact assessment can be used to provide ex-ante and ex-post evaluations of river basin management plans. This allows quantification of costs of river basin management options, benefits from these investments, expected return on investments and the need to quantify the full cost of infrastructure (river basin works such as irrigation infrastructure, waste water treatment plants) and management actions (for example, soil conservation measures, vegetation retention, non-point source pollution management).

Musgrave and Sinden (1988) provided one of the first conceptualisations of funding catchment management, using the metaphor of a condominium. Here, a common 'title'—the body corporate—provides a joint ownership mechanisms for managing the 'commons' of upkeep on the condominium. Regular contributions by members of the body corporate fund maintenance operations. A similar argument is put forward for managing catchment natural resource assets, which though not commonly owned, as most are in private ownership, still have a common net impact of providing positive benefits to catchment (the outputs of industrial and agricultural activities, tertiary services outputs) and disbenefits (deterioration of water quality, soil erosion).

More recently, Marshall (1997) suggest that,

> A resource economics approach uses 'economic instruments' to respecify private property rights so that externalities are internalised. But only if benefits are not likely to be outweighed by costs of government failure. Such instruments (e.g. tradeable permits) are typically preferred to the standard-setting that is normally integral to planning approaches.
>
> (p. 5).

This approach regards common property regimes, such as river basins, as difficult places to enact a common property economic regime (CPR) with many economists disfavouring the use of a CPR as a means of creating an economic tool for its management. Marshall goes on to argue that in IRBM (or at least smaller scale, sub-basin management), common rights are often broken down into individual rights by catchment management committees (put in place by a mix of voluntarism and government incentive) and rely on mutual reassurance and goodwill to enact catchment management. However, this approach has a limited life as good will is often tested when resource managers' practices negatively impact others.

The challenge remains how to address these emerging economic issues and present robust models for financing common-pool resource management in a river basin.

7.5.5 Cost-sharing arrangements

One economic instrument which shows promise in IRBM is the use of cost-sharing arrangements. Fargher (1997) articulated the opportunity for cost-sharing in paying for river basin management in a clearly defined continuum of funding opportunities (Figure 7.2). Here, the continuum is scaled according to the degree of contributions between the polluter and those being polluted.

Cost-sharing is a mechanism where the financing of river basin management practices, options, programmes and initiatives is spread between beneficiaries.

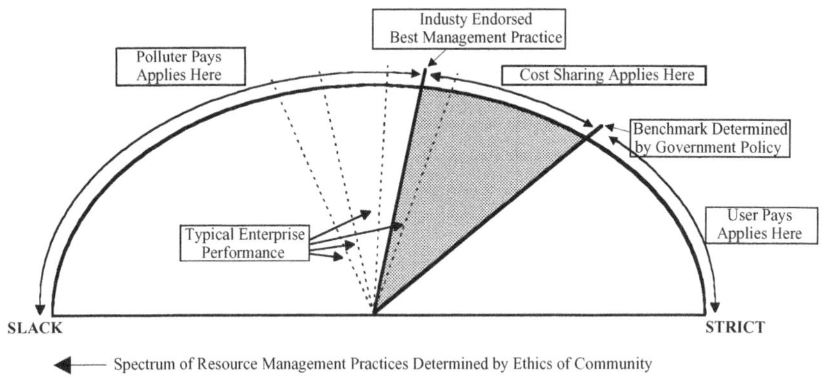

Figure 7.2 Relationship between community ethics and government policy, and the place of cost-sharing.
Source: Fargher 1997.

In Figure 7.2, the maximum extent of cost-sharing is determined according to an upper limit determined by government (who considers the public interest and 'good') beyond which the user will pay—for example, as a cost of water supply or an 'environmental' cost (e.g. a tax) for the provision of environmental management services. This can be charged as a catchment levy or a water user supply cost. The lower limit is determined where the cost directly transfers to polluters (such as in a fine or pollution discharge cost). Between these two limits there is a broad range of opportunities for industries, cities and agricultural businesses to appropriate some of the cost of providing environmental services through adoption of best management practices, at different levels of environmental management performance—quality benchmarks.

A component of cost-sharing is the determination of beneficiary pays. This principle suggests, in its mildest form, that all beneficiaries meet some portion of the cost and that together the beneficiaries cover all the costs. While in a more rigid interpretation, costs are distributed in the precise proportion of their shares of estimated benefits, but this needs understanding of beneficiaries' ability to pay, costs of collection, and determination of fault (Marsden 1996). These cost-sharing principles have been applied to basin management in the Murray–Darling Basin with some success in district LWMPs and throughout the basin at a larger scale (Murray–Darling Basin Commission 1996; Oram and Dumsday 1996; Marshall 1997; Curtis and Lockwood 2000; Marshall 2002) and in the coastal river valleys of New South Wales where the use of market-based instruments and 'green-offsets' were promoted (Healthy Rivers Commission of New South Wales 2003).

7.5.6 Pollution trading permits and the polluter pays principle

Pollution trading permits are an economic management tool which has been defined and has had some experimental use. A permit to trade pollution is defined as,

> Individual polluters can be allowed the right to buy and sell quotas of emissions subject to an overall upper quota on total emissions. Nutrient trading is a potentially useful instrument to improve water quality.
>
> (Global Water Partnership 2002).

The value of a tradeable permit to river basin management is that it creates a property right for pollutants which can be bought and sold. These pollutants can be traded in a market place in that they allow polluters to buy assimilative capacity in the environment. Robinson and Ryan (2002) considered five requirements be addressed before establishing a market for trading water pollution:

Nature of emission source: Tradeable permits are mainly useful when dealing with point source pollution where it is feasible to measure discharge from individual sites, such as nutrients from sewage treatment plants. They are inappropriate for non-point source problems such as sediment loads from stormwater.

Determining cap and participants: Greater numbers of participants reduce the marginal cost of abatement and hence increase economic efficiency.

Defining the permit: Defining the scope of a permit (pollutant type and circumstances of generation) is essentially a scientific exercise whereas the size of the permit (i.e. 1 kg or 1 tonne, etc.) is a question of economic and administrative efficiency. In any case, the permit should be defined with enough legal certainty that they can confidently be traded.

Method of permit allocation: Initial allocation of permits is a critical issue. Despite a theoretical preference for an auction approach, 'grandfathering' of the initial permits has been applied to virtually all applications in practice.

Administrative structure, monitoring and enforcement: Tradeable permit systems are often perceived as requiring costly monitoring though any effective system would require monitoring. A prerequisite for the successful implementation of tradeable permits is a credible system for their enforcement.

Source:http://www.coastal.crc.org.au/planning_compendium/paper_robinson/ei_discharge .htm, accessed December 2004.

Trading schemes initially appear promising as an IRBM tool but limited experience suggests they may remain a theoretical construct for the time being. Trading schemes can be intensive in terms of information and enforcement, hence costly to administer; the high transaction costs of certain markets may outweigh their benefits (Global Water Partnership 2002).

An alternative to a trading-based system is a regulatory system in which charges are levied against those who discharge pollutants into a river or groundwater system in a basin. This is more common than the trading system and has had more use. Regulatory tools are discussed in Section 7.5.8.

7.5.7 Covenants of mutual obligation

Covenants of mutual obligation provide an opportunity for implementing natural resources management arrangements in IRBM. Covenants are reciprocal agreements for action between stakeholders in a river basin. A covenant is often defined as a land title or a constraint on landuse over a parcel of land. But this is not the meaning here, rather it refers to the Biblical concept of promise and lasting agreement.

IRBM covenants include several elements:

- A vision statement of the desired future for land and water resources management for a specified period
- Clear identification and specification of water rights
- Identification of cost-sharing arrangements to share river basin management expenses
- Clear specification of river basin works and targeted actions, and who is responsible for each,

- Contractual agreements between stakeholders including government departments, industry organisations, water user groups and RBOs to undertake actions
- Promises of commitment to the process by participating entities
- An external review process to audit outcomes.

There are several factors which need to be addressed to implement covenants of mutual obligation:

- Building and maintaining strong leadership of the 'engaged' communities;
- Ensuring membership of catchment management organisations is on a skills basis, not on a representative basis;
- Developing trust between all stakeholder groups to agree on each other's mutual obligations;
- Ensuring a sustained funding base;
- Designing a clearer definition of property rights;
- Understanding the reasons why landowners are unable and unwilling to adopt sustainable land and water management practices;
- Use of well trained facilitators;
- Using support tools such as interactive River Basin Information Systems (see Chapter 6).

There is the need for a government agency to establish mechanisms to create RBOs at the interstate or sub-state level which can enact basin covenants. One way to do this is to establish three-way funding mechanisms between the national, state and local governments which will facilitate:

- the formation of an RBO and programmes to cost-share river basin management with private sector interests.
- leadership training and further research in IRBM with the establishment of a national training and research institute
- the implementation of river basin information systems to facilitate information exchange amongst river basin managers.

7.5.8 Rules and regulations

'Rules', in the context of institutional arrangements for IRBM, define what a river basin management organisation must do, may do and must not do (Scott 1995). They also specify the area over which an RBO may work. Rules for international RBOs are different to those of national RBOs. An example of an international 'rule' is the Law of the Non-navigational Uses of International Water Courses as discussed in Section 7.5.2. A national 'rule' could be the national water policy of a country which specifies roles and responsibilities of RBOs within the nation's boundaries.

A 'regulation', in the context of institutional arrangements for IRBM, is defined as the range of laws which enable river basin management. Such legal specifications are more common to nations rather than between nations. Millington (2004) suggests that IRBM,

> must be supported by a legislative framework that clearly stipulates the role and powers of any new basin organizations and how they interact with existing agencies, departments and bureaus. These requirements are basically the same irrespective of whether they are included within a special-purpose 'RBM' act, or within an existing piece of legislation such as a water resources law.
>
> The legislation can be coordinating/planning/monitoring frameworks that deals with how basin organizations are to be created and to operate.
>
> (Briefing Note 2. Creating a Basin Organization, p. 2).

Such legislation provides the specification of the role of an RBO, its geographical scope, membership of committees, roles, responsibilities, monitoring procedures, river basin planning processes, information needs, river basin plan contents and reporting mechanisms. This provides a strong foundation for IRBM and is to be preferred to high-level decrees by ministers of government or presidents, or ad hoc, voluntary arrangements. It sets a secure basis for present and future river basin management operations. One operational role that legislation can perform is to specify how river basin decision-makers participate under a contractual arrangements, rather than ad-hoc, voluntary arrangements which are subject to political influence and participant burnout.

The downside to regulatory frameworks is the inability of governments to implement them. For example, within the State of Andhra Pradesh, India, there are several pieces of legislation which were passed by governments, yet the administrations during and since their decree have never had the resources to implement them. This is why caution is needed in establishing a regulatory-based approach as the preferred governance tool for IRBM. Laws provide administrative, fiscal and regulatory stability, but they are more a feature of stable, wealthy democracies than those countries with limited financial resources and administered by wealthy elites or despotic governments. As mentioned earlier in this book, it is more likely that IRBM will be implemented if democratic processes are at work in the governance of society.

There are a wide range of regulatory tools which can be used for natural resources management at the basin and sub-basin level. These can be recognised in any setting and included within a river basin organisation's mandate, but their application must be crafted to suit the context. Global Water Partnership (2002) lists these regulatory tools as:

Direct regulations, whereby government bodies or independent regulatory agencies establish laws, rules or standards which water and land users and water service providers are required to follow. This is often known as command and control regulation. Such regulations might, for example, include the specification of drinking water quality standards, controls over landuse and development within catchments and flood plains, controls over the quantity and timing of private water abstractions, and controls over the quantity, quality and timing of waste discharges into the water environment.

Economic or market regulation—economic instruments such as unit pricing, marketable rights or subsidies are employed instead of or in conjunction with direct regulations to influence water or land using behaviour.

Self regulation—professional bodies, industry groups or community groups establish their own rules of conduct and mechanisms to ensure compliance. Governments may still have an important role, however, in allowing self-regulating systems to operate, in encouraging, enabling and building regulatory capacity and in providing vital information.

Social regulation—this involves changing water use behaviour through persuasion, information and education.

Source: Tool C6. Regulatory Instruments

The use of regulatory tools is best conceptualised in a management arrangement which separates the roles of the regulator from those of the resource manager and service provider. This facilitates improved accountability and streamlines functions of the water sector, within which RBOs operate. Millington (1999) defines each:

Regulator/standard setter

Develops and implements a financial/economic or pricing regulatory regime.

Develops water quality and other natural resource objectives, standards or guidelines.

Audits the performance of the water sector as to compliance with standards.

The natural resources manager

Undertakes strategic water assessments.

Develops policies and strategies to comply with national objectives and with standards set by regulator. Also develops and oversights a strategic water research programme.

Develops legislation to support regulatory standards and policies.

Plans and allocates water.

Manages quantity and quality for surface water and groundwater.

Supports inter-agency and community driven basin coordination.

Develops water sector capacity building programmes.

Promotes public participation and water awareness.

The operator/service provider

Builds and operates water supply, sewerage, drainage and irrigation systems.

Maintains infrastructure.

Provides technical advice and assistance to others.

Charges others for services provided.

Operates under some form of contract(s), usually to the regulator for operating rights and to the resource manager for utilisation of the water resource.

The establishment of this three-fold enabling environment creates a stable framework for IRBM.

7.5.9 The role of local government in IRBM

Local government has strong resource allocation powers and a clear local mandate and jurisdiction in environmental management, particularly in highly developed economies and democracies. These powers can be harnessed for IRBM to enact local sub-basin management.

Local government powers can be used within an Integrated River Basin Management Plan and a Land and Water Management Plan (LWMP). The latter is the local component, a sub-basin plan. This hierarchical approach suggests the need for congruence between plans and planning processes at all scales. An example of an LWMP was presented in Chapter 4.

Progress has been made in Ontario, Canada, using a procedure named Watershed-Based Source Protection Planning (Ontario Ministry of Environment and Energy 1993) to enact natural resources management at the sub-basin scale. This involved the use of a hierarchical planning process from Watershed Plans to Subwatershed Plans to Site Management Plans (Figure 7.3). The last includes local specification of best management practices, permits and construction approvals and the need for environmental assessments. Site Management Plans are designed to be compatible with Subwatershed Plans (which detail subwatershed targets, goals and monitoring requirements) and broader Watershed Plans (policies on ecological integrity and carrying capacity, water quality and quantity management). The Ontario government also provided a guide for municipalities to enact

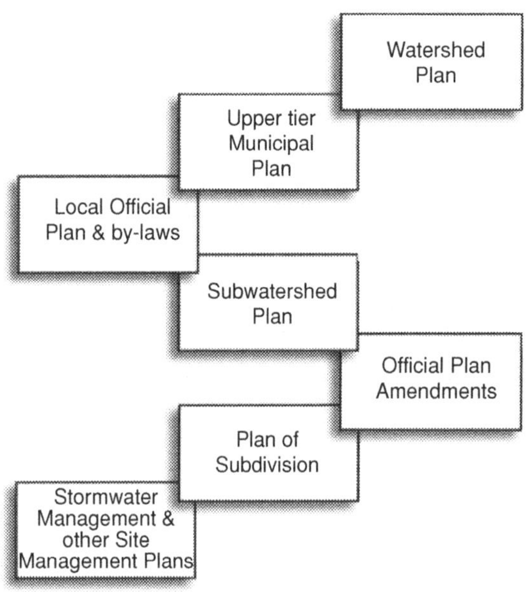

Figure 7.3 Ontario's hierarchical arrangements of watershed plans, subwatershed plans and stormwater and other site management plans and municipal planning documents. *Source:* Ontario Ministry of Environment and Energy (1993).

watershed management in municipal planning documents. This is done by coordinating state and local agency programmes. The approach is based on the review of development applications, so that new developments are reviewed for congruence with broader ecosystem quality, water quality and quantity goals. The strength of this approach is to set targets and constraints (for example flood susceptibility) that are subsequently used to evaluate how well the subwatershed plan meets watershed plan objectives.

In a second example, the role of local government was reviewed in the former Hawkesbury-Nepean Catchment Management Trust in Australia (AACM International 1997) and a review of the functions of that same Trust (Burton, Hooper and Junor 1995). The Hawkesbury-Nepean basin (22,000 km^2 in size) is located on the periphery of metropolitan Sydney, and is experiencing rapid urbanisation and transformation from a primarily rural landscape. The studies documented this RBO's role in local government planning. The Trust was found to have a valuable role in river basin management, in its ability to:

- Advise municipal councils on the overall impacts of new large-scale housing developments in the context of overall basin management
- Provide State of the Environment advice to Local Government and thus continually influences on local planning decisions
- Advise on the coordination of natural resources management throughout the basin, including local government's role
- Assessing the investment of councils in the basin in environmental protection and catchment management—found to be about $A200 million each year
- Be an advocate for Basin issues within a municipality or shire[5] where there was no such voice due to local government's reluctance to speak beyond its boundaries.

The first of these studies, AACM International (1997), recommended a procedure for defining investment and cost-sharing mechanisms between the Trust and local councils (the mechanisms were discussed in Section 7.5.5). The Trust had the governance role of providing:

- Strategies for managing common priorities
- Frameworks for common responsibilities
- Coordination of related activities
- Research into cross-boundary issues
- Sharing the cost of regional resource management
- Attracting external investors in regional natural resource management to the river basin.

The ability to enact IRBM depends on many factors, a theme we have seen throughout this chapter. One of the skills is the ability of natural resources managers

[5] A shire in Australia is a local government administrative district, similar to a county in the United Kingdom and United States and a district in India.

to broker agreements between local government stakeholders in a river basin context. The challenge is to work towards agreed programmes of action rather than actions which cause separate uncoordinated decision-making and conflict. Long-term river basin management planning can be commenced simultaneously with the goal of developing an agreed, cost-shared plan of action, and a review of initial best management options. In short, there is a matrix by which investment, targeted action and coordination mechanisms can be the hallmarks of good river basin governance by local governments, and they provide local leadership in natural resources management.

7.6 MEASURING RIVER BASIN GOVERNANCE PERFORMANCE

In this chapter, we discussed the social and institutional capacity of river basin organisations and river basin initiatives as a means to implement an IRBM. But how effective have river basin organisations been in implementing this integrated approach? Effectiveness measurement involves developing two sets of indicators:

- Performance indicators of existing river basin organisations or existing institutional arrangements for river basin governance—how well does the organisation/institution do? This is discussed in Section 7.6.1.
- Basin ecosystem sustainability indicators—how well does the natural environment respond? Numerous highly developed basin indicator procedures have been developed to measure the ecosystem health of basins.

As the focus of this book is governance, this section develops a preliminary set of performance measures while basin sustainability indicators are reported elsewhere. The critical questions are how can the performance of a river basin organisation be measured? What are the performance measures used to demonstrate improved governance? Does an integrated approach to basin governance show improvements compared to single-sector basin management?

7.6.1 Measuring RBO Performance

'Good' river basin management was described in Table 7.1 by 34 attributes assembled into three groups: an enabling environment, the institutional framework and management tools. These groups form 'good governance' factors, in that they assist implementation of IRBM by RBOs. Table 7.5 provides a preliminary set of factors of good governance. Diagnostic indicators will be developed from these factors and is the focus of current research by the author.

This approach to measuring RBO performance was developed from empirical evidence of successes and failures in IRBM (Burton, Hooper and Junor 1995; Barrow 1998; Hooper and Moraitis 1998; Bellamy et al. 1999; Hooper, McDonald

Table 7.5 Components used to develop diagnostic/performance indicators for river basin management organisations

Components influencing good governance	Description
Policy	Evidence and effective use of supporting land and water policy
Legislation	Strong and comprehensive, but flexible legislation, regulations, decrees, etc., which ensure 'fairness' in basin-wide decisions and a process of accountability
Finance	Tools and mechanisms used to fund IRBM
Institutional stability	Stable institutional framework exists which overcomes fragmentation and overlap of responsibilities
Cross-sectoral linkages	Degree of high level, cross-sectoral links between health, water, population and economic development, which means agencies do not find singular solutions but look at impacts and improvements across the spectrum of natural resources, and the development of regional (basin scale) natural resources management policies
Goal shift	Move from a pure resource exploitation ethic to incorporate environmental management and incorporate sustainable and appropriate development goals
Legal setting	Appropriateness and degree of enforcement of the legal and jurisdictional setting
RBO role specificity	Existence and use of river basin organisation (RBO) roles, responsibilities and functions
Jurisdiction over an informal water sector	Lack of regulation of water use when there are vast numbers of small-scale users and ground water pumpers who are not linked with public institutions
Realistic functions	Degree to which RBO roles, responsibilities and functions reflect realities of existing conditions
Evidence of coordination mechanisms to enact integrated management	Avoid bias in monitoring, planning and management through coordination of a range of state, federal, commercial and private NGO bodies
Specificity of the problem domain: scope, scale and boundary identification and realistic goal formulation	Need for well-defined objectives, mutually beneficial and desirable goals, resource development forms part of a long-term integrated basin management plan; awareness of constraints on development in basin
	Realistic and informed understanding of what are the feasible options
	Definition of the scope of the problem-shed, range of issues, environmental policies and management activities; a clear boundary of the edge of the problem domain is established

(Continued)

Table 7.5 Components used to develop diagnostic/performance indicators for river basin management organisations (*Continued*)

Components influencing good governance	Description
Characteristics of organisational style—degree to which they address conflict resolution, coordination and cooperation	Satisfactory organisational structures which allow cross-sectoral planning and management Focus on coordination and advisory roles; more than one body forms the basin entity—focussed on oversight, management and planning
Rules governing structure: position and boundary of entities rules are clearly defined	Specification of the entities involved in RBM and their roles in decision settings; 'rules' of participation specify roles of participation and membership and exiting decision settings
Rules governing process: authority rules	Rules are defined here for the array of coordination activities (who is involved), how binding or permissive is the coordination (what can be done) and on what basis is the involvement (law, policy, informal agreement)
Rules governing process: information rules	The content of the information used by participant is specified, the form of the information and the timing of information exchange
Rules governing process: decision rules	Specification of how decisions are made (consensus, voting, etc.)
Financing tools	Degree to which financial arrangements are transparent, ongoing and of enough quantity to make a difference
Cost-sharing tools	Mechanisms for cost-sharing river basins management programmes and practices
Human resources capacity building	Well-trained staff with capacity to work in teams and plan across sectors and disciplines
Dialogue	Use of dialogue as a tool to make decisions on preferred management options
Monitoring	Use of a strong knowledge base that derives from a good, uniform and comprehensive data network, systems and models for analysis, and that allows 'knowledgeable' natural resources/water management policies and strategies to be developed and implemented and which links to decision system of the basin
Public sector role	Specification of the role of state and local agencies to satisfactorily implement development and resource management activities
Private sector participation	Specification of private sector involvement and links to basin decision systems
Community participation	Specification and use of community participation mechanisms Use of strong community awareness and participation processes—to enhance greater farmer ownership of basin scale plans of action

Table 7.5 Components used to develop diagnostic/performance indicators for river basin management organisations (*Continued*)

Components influencing good governance	Description
Accountability mechanisms	Accountability mechanisms of RBO to higher authorities and citizens 'Policing' by an independent body (or bodies) with enough authority to insist on improvements
Adaptive management	Use of a 'learn by doing' and 'development facilitator' approach. Use of a flexible, adaptive management decision system, reacting to new research and understanding and goal shifting
Information use	Establishment and use of knowledge base which specifies courses of action for basin sub-units
Culture of research-knowledge links	Existence of research system (and data collection and analysis) and its spatial presentation to inform decision-making (GIS) Provision of data, monitoring and understanding of the basin structure and function and resource activities
Role of regional and local government	Degree of local and regional cross-sectoral integration
Vertical management linkages	Degree of hierarchical integration of decision system
Informal sector role and influence	Extent of jurisdiction over an informal water sector
Prioritisation of efforts	Ability to address critical problems first: water scarcity, flooding, droughts, e.g. very large and rapidly growing populations depend on a limited natural resource. The challenge is to get more crop, cash and jobs for each drop
Productivity efficiency	Evidence of increasing the productivity of water diverted from rivers is less important than being able to capture water more effectively in the soil profile
Organisation efficiency	Use of best management practices within the RBO
Push towards sectoral best practices	Use of local best management options in industry, urban planning and agricultural practices
Information	Information use: degree to which information is accessible, appropriate, equitable, affordable and integrated

ISO = International Standards Association
EMAS = Environmental Management and Audit Scheme, the European Union's voluntary, eco-auditing and management system.

and Mitchell 1999; Millington 1999; Syme and Butterworth 1999; Bellamy and Dale 2000; Walmsley et al. 2001; Pitman 2002; Shah, Molden and Sakthivadivel 2003b). Indicators will be developed to assess the degree to which mechanisms have been implemented to address current problems in different basin settings.

The result of implementation will be reflected, at varying times and spatial scales, by a range of sustainability indicators in a catchment. Any river basin can be assessed for the degree to which these factors have shown demonstrable change. The challenge is to demonstrate the degree and extent to which these changes can be attributed to the RBO organisation, specifically it is the effectiveness of its river basin management plans or other key functional tasks.

Table 7.5 was designed for multiple contexts: for large and small basins, for high-rainfall and low-rainfall environments, developing and developed country scenarios. The critical issue is the extent to which each performance indicator can be applied, once developed. The challenge is to develop an indicator which measures the presence, use and effectiveness of each component/element. That is, for each indicator there are probably at least three measures. However, such a fine diagnostic will be limited by data availability.

In any setting, there will be considerably more work required to develop these indicators to match specific river basins and to identify data sets.

7.7 SUMMARY AND FUTURE DIRECTIONS

In this chapter, we have seen how the creation and ongoing operation of IRBM is a function of many social factors and institutional arrangements. Effective IRBM varies according to how these factors can be harnessed to suit local needs. It is important to recognise the fundamental differences between IRBM in countries of the South and North—what works in one place is not necessarily transferable to another.

This discussion has highlighted what some may consider 'first principles' and has reflected on selected experiences. Considerable work is required in more applications of the social and institutional environment, and to develop measures which clearly demonstrate that efforts have been worthwhile.

This book has raised many questions about the complexity of IRBM, issues which will be resolved as more experiences produce more 'lessons learned'. The future of many river basins today will be decided by the management decisions of our current and the next generation of basin managers and numerous individual decision-makers. It is hoped that this book has provided insight into how these stewards can be equipped to make better, more informed decisions.

7.8 REFERENCES

AACM International. 1997, Local Government and Catchment Management. Draft Report for the Hawkesbury-Nepean Catchment Management Trust. Adelaide, AACM International.

AACM International & Centre for Water Policy Research. 1995a, Enhancing the Effectiveness of Catchment Management Planning. Annex B: Activity Review. Adelaide, AACM with the Centre for Water Policy Research.

AACM International & Centre for Water Policy Research. 1995b, Enhancing the Effectiveness of Catchment Management Planning. Final Report. Final Report for the Department of Primary Industries and Energy. Adelaide, AACM International Pty. Ltd.

Alaerts, G. 2001, River Basin Management: Developing Institutions—A Framework. Fifth River Basin Workshop: Latin American and African Examples. Report from 5th River Basin Workshop: Latin American and African cases.

Allee, D. J., Apener, H. R., & Andrews, W. H. 1975, Basin Governance. Cornell Agricultural Economics Staff Paper. [No. 1]. Ithaca, NY, USA, Cornell University.

Anonymous. 2000, Towards People Oriented River Basin Management: An NGO Vision. Paper presented to The First World Water Forum. Publisher unknown.

Anonymous. 2002, IWRM and Basin Management Theme. Final Statement. World Water Council.

Australian Manufacturing Council. 1992, The Environmental Challenge: Best Practice Environmental Management. Melbourne, Australian Manufacturing Council Secretariat.

Barham, E. 2001, Ecological boundaries as community boundaries: The politics of watersheds. Society and Natural Resources 14[3], 181–191.

Barrow, C. J. 1998, River basin development planning and management: A critical review. World Development 26[1], 171–186.

Bellamy, J. A. & Dale, A. 2000, Evaluation of the Central Highlands Regional Resources Use Planning Project: A Synthesis of Findings. Brisbane, Australia, CSIRO Sustainable Ecosystems.

Bellamy, J. A., McDonald, G. T., Syme, G. J., & Butterworth, J. E. 1999, Evaluating integrated resource management. Society and Natural Resources 12, 337–353.

Bellamy, J. A., Walker, D. H., McDonald, G. T., & Syme, G. J. 2001, A systems approach to the evaluation of natural resource management initiatives. Journal of Environmental Management 63[4], 407–423.

Born, S. M. 2000, Emerging Trends and Innovations in Institutional Arrangements. International Riversymposium, Proceedings. Brisbane, International Riverfestival.

Born, S. M. & Genskow, K. D. 1999, Exploring the "Watershed Approach"—Critical Dimensions of State-Local Partnerships: The Four Corners Watershed Innovators Initiative. Final Report. Madison, WI, USA, Department of Urban and Regional Planning, University of Wisconsin-Madison.

Born, S. M. & Genskow, K. D. 2000, The Watershed Approach: An Empirical Assessment of Innovation in Environmental Management. 7. Washington, DC, National Academy of Public Administration.

Born, S. M. & Genskow, K. D. 2001, Towards Understanding New Watershed Initiatives: A Report from the Madison Watershed Workshop. 1–23. Madison, Wisconsin, University of Wisconsin-Madison Cooperative Extension Service.

Bruins, R. J. F., Heberling, M. T., & Maddock, T. A. 2005, Conceptual Approach for Integrated Watershed Management. [9]. Chapter 9 (pp. 185–226) in Economics and Ecological Risk Assessment: Applications to Watershed Management. R. J. F. Bruins & M. T. Heberling. Eds. CRC Press, Boca Raton, Florida.

Burton, J. R., Hooper, B. P., & Junor, R. S. 1995, Review of the Hawkesbury-Nepean Catchment Management Trust. Armidale, Centre for Water Policy Research.

Cameron, J. I. 1997, Applying socio-ecological economics: A case study of contingent valuation and integrated catchment management. Ecological Economics 23[2], 155–165.

Cantwell, M. & Day, J. C. 1998, Citizen-initiated river basin planning: The Salmon River Watershed example. Environments 25[2–3], 80–90.

Chave, P. 2001, The EU Water Framework Directive. An Introduction. London, IWA Publishing.

Correira, F. N. & da Silva, J. E. 1999, International framework for the management of transboundary water resources. Water International 24[2], 86–94.

Cosgrove, W. J. & Rijsberman, F. R. 1999, World Water Vision. Making Water Everybody's Business. London, UK, Earthscan Publications Ltd.

Curtis, A. & Lockwood, M. 2000, Landcare and catchment management in Australia: Lessons for state-sponsored community participation. Society and Natural Resources 13[1], 61–73.

Delbridge, A. (ed.). 1982, The Concise Macquarie Dictionary. Lane Cove, Sydney, Doubleday.

Falkenmark, M., Gottschalk, L., Lundqvist, J., & Wouters, P. 2004, Towards integrated catchment management: Increasing the dialogue between scientists, policy-makers and stakeholders. Water Resources Development 20[3], 297–309.

Fargher, J. D. 1997, Local Government and Catchment Management. Draft Report for the Hawkesbury–Nepean Catchment Management Trust. Adelaide, AACM International.

Global Water Partnership. 2002, Integrated Water Resources Management ToolBox. Stockholm, Global Water Partnership.

Global Water Partnership. 2003, Effective Water Governance: Learning from the Dialogues. Report from "3rd World Water Forum" Kyoto, Japan.

Global Water Partnership Technical Advisory Committee. 2000, Integrated Water Resources Management. TEC Paper. [No. 4]. Stockholm, Global Water Partnership.

Green, C. 2004, Integrated Water Resource Management as Conversation. Enfield, London, Flood Hazard Research Centre, Middlesex University.

Harmsworth, G. 1998, Maori values for catchment management. Catchment connections 3, 12. Landcare Research. Palmerston North, New Zealand.

Havekes, H., Koemans, F., Lazaroms, R., Poos, D., & Uitjterlinde, R. 2004, Water Governance: The Dutch Water Board Model. The Hague, Dutch Association of Water Boards.

Healthy Rivers Commission of New South Wales. 2003, Healthy Rivers for Tomorrow. Sydney, Healthy Rivers Commission of New South Wales.

Hooper, B. P., McDonald, G. T., & Mitchell, B. 1999, Facilitating integrated resource and environmental management: Australian and Canadian perspectives. Journal of Environmental Planning and Management 42[5], 746–766.

Hooper, B. P. & Moraitis, E. 1998, Strengthening Resource and Environmental Management in the Tweed Catchment. Towards a Template for New South Wales Coastal River Valleys, Estuaries and the Coastal Zone. Brisbane, Department of Geographical Sciences and Planning, The University of Queensland.

Imperial, M. T. 2004, Collaboration and performance management in network settings: Lessons from three watershed governance efforts. Lawrence, P. & Kamensky, J. M. (eds.). Washington, DC, IBM Center for The Business of Government. Collaboration Series.

Lamoree, G. & Nilsson, A. 2000, A process approach to the establishment of international river basin management in Southern Africa. Physics and Chemistry of the Earth Part B: Hydrology Oceans and Atmosphere 25[3], 315–323.

Lant, C. 2003, Watershed governance in the United States: The challenges ahead. Water Resources Update 126, 21–28.

Leach, W. D. & Pelkey, N. W. 2001, Making watershed partnerships work: A review of the empirical literature. Journal of Water Resources Planning and Management 127[6], 378–385.

Le Moigne, G., Barghouti, S., Feder, G., Garbus, L., & Mei, X. 1992, Country experiences with water resources management: Economic, Institutional, Technological and Environmental Issues. 175. Washington, DC, USA, World Bank. World Bank Technical Paper.

Le Moigne, G., Subramaniyan, A., Mei, X., & Giltner, S. 1994, A Guide to the Formulation of Water Resources Strategy. 263. Washington, DC, World Bank. World Bank Technical Paper.

Letey, J. 1999, Science and policy in integrated watershed management: A case study. Journal of the American Water Resources Association 35[3], 603–607.

Lewicki, C. 2001, Supporting community-based watershed efforts. Water Resources Update 119[February], 9–14.

Margerum, R. D. 1995, Integrated Management for the Wisconsin Department of Natural Resources: An Overview. Wisconsin, Wisconsin Department of Natural Resources. Bureau of Water Resources.

Marsden, J. 1996, Cost-Sharing—The Murray Plans. Sustainable Management of Natural Resources: Who Benefits and Who Should Pay? [No. 6]. Canberra, Land and Water Resources Research and Development Corporation. Occasional Paper.

Marshall, G. R. 1997, Integrated Catchment Management: A Road Paved with Good Intentions. Seminar to the Agricultural and resource Economics Group, Faculty of Agriculture, University of Western Australia. Armidale, Department of Economics, University of New England.

Marshall, G. R. 2002, Institutionalising cost sharing for catchment management: Lessons from land and water management planning in Australia. Water Science and Technology 45[11], 101–111.

McCool, S. F. & Guthrie K. 2001, Mapping the dimensions of successful public participation in messy natural resources management situations. Society and Natural Resources 14[4], 309–323.

Millington, P. 1999, River Basin Management—Its Role in Major Water Infrastructure Projects (Draft). Cape Town, South Africa, World Commission on Dams Secretariat. World Commission on Dams Thematic Reviews. Institutional Processes V.3 River Basins—Institutional Frameworks and Management Options.

Millington, P. 2004, Integrated River Basin Management—From Concepts to Good Practice, Briefing Notes to the World Bank (Draft).

Mullen, M. W. & Allison, B. E. 1999, Stakeholder involvement and social capital: Keys to watershed management success in Alabama. Journal of the American Water Resources Association 35[3], 655–662.

Murray–Darling Basin Commission. 1996, Cost-Sharing for On-Ground Works. Canberra, Murray–Darling Basin Commission.

Musgrave, W. F. & Sinden, J. A. 1988, An Economic View of Integrated Catchment Management. Workshop on Integrated Catchment Management. Canberra, Australian Water Resources Council.

Oliver, P. 2001, What makes catchment management groups "tick"? Water Science and Technology 43[9], 263–272.

Ontario Ministry of Environment and Energy. 1993, Subwatershed Planning. Toronto, Queen's Printer of Ontario.

Oram, D. & Dumsday, R. 1996, Socio-Economic Analyses of Dryland Salinity Control in the Murray–Darling Basin. Project Status Report. Gunnedah, Liverpool Plains Land Management Committee.

Parkes, M. & Panelli, R. 2001, Integrating catchment, ecosystems and community health: The value of participatory action research. Ecosystem Health 7[2], 85–106.

Pitman, G. K. 2002, Bridging Troubled Waters: Assessing the World Bank Water Resources Strategy. Washington, DC, USA, World Bank.

Robinson, J. & Ryan, S. 2002, A Review of Economic Instruments for Environmental Management in Queensland. 3.5.3. Brisbane, Coastal Cooperative Research Centre. Research Products.

Rogers, K., Roux, D., & Biggs, H. 2000, Challenges for catchment management agencies: Lessons from bureaucracies, business and resource management. Water South Africa 26[4], 505–511.

Rogers, P. & Hall, A. W. 2003, Effective Water Governance. Stockholm, Global Water Partnership.

Saeijs, H. L. F. & Turkstra, E. 1994, Plea for sustainable development of European river basins. European Water Pollution Control 4[3], 16–28.

Scott, R. W. 1995, Institutions and Organizations. London, Sage Publications.

Shah, T., Molden, D., & Sakthivadivel, R. 2003a, Limits to Leapfrogging: Issues in Transposing Successful River Basin Institutions in the Developing World. Colombo, Sri Lanka, IWMI-TATA Water Policy Program. Water Policy Briefing.

Shah, T., Molden, D., & Sakthivadivel, R. 2003b, Limits to Leapfrogging: Issue in Transposing Successful River Basin Institutions in the Developing World. Colombo, Sri Lanka, IWMI-TATA Water Policy Program. Water Policy Briefing.

Shuval, H. 1994, Proposals for the integrated management of the shared transboundary water resources of the Jordan River Basin. Water Science and Technology 30[5], 187–193.

Syme, G. J. & Butterworth, J. E. 1999, SHEF: Self-Help Evaluation Framework for Integrated Catchment Management. Final Project Report. Project CWW22. Canberra, Land and Water Resources Research and Development Corporation.

Van Steenbergen, F. & Shah, T. 2002, Rules rather than rights: Self regulation in intensively used groundwater systems in Intensive Use of Groundwater: Challenges & Opportunities. Llamas, M. R., Custodio, E. (eds.) [12], 241–256.

Walker, S. 2001a, Balancing social, economic and environmental pressures through integrated river basin management in the Cairngorm Mountains of Northeast Scotland. 45–50. IAHS AISH Publication [272].

Walker, S. 2001b, Opportunities for balancing conflicting economic, social and environmental pressures on river basins through an integrated approach with stakeholder involvement. IAHS AISH Publication [268], 175–182.

Walmsley, J., Carden, M., Revenga, C., Sagona, F., & Smith, M. 2001, Indicators of sustainable development for catchment management in South Africa—Review of indicators from around the world. Water South Africa 27[4], 539–550.

Waterman, R. J., Peters, T., & Phillips, J. R. 1980, Structure is not Organisation. Business Horizons 23[3], 14–26.

Wolman, A. 1981, Some reflections on river basin management. Water Science and Technology 13[3], 1–6.

Annex: international river basins

www.transboundarywaters.orst.edu

The author acknowledges the assistance and the data provided from the TRANS-BOUNDARY FRESHWATER DISPUTE DATABASE, directed by Dr. Aaron T. Wolf, Oregon State University, in the compilation of the Annex to this book which provides data on international river basins for the continents of Africa, Asia, Europe, North America and South America.

Basin name	Total area of basin (sq. km) (1)	Country name	Area of country in basin (sq. km)	Percent area of country in basin (%)
Akpa (2)	4900	Cameroon	3000	61.65
		Nigeria	1900	38.17
Atui (3)	32600	Mauritania	20500	62.91
		Western Sahara	11200	34.24
Awash	154900	Ethiopia	143700	92.74
		Djibouti	11000	7.09
		Somalia	300	0.16
Baraka	66200	Eritrea	41500	62.57
		Sudan	24800	37.43
Benito/Ntem	45100	Cameroon	18900	41.87
		Equatorial Guinea	15400	34.11
		Gabon	10800	23.86
Bia	11100	Ghana	6400	57.58
		Ivory Coast	4500	40.28
Buzi	27700	Mozambique	24500	88.35
		Zimbabwe	3200	11.65

Basin name	Total area of basin (sq. km) (1)	Country name	Area of country in basin (sq. km)	Percent area of country in basin (%)
Cavally	30600	Ivory Coast	16600	54.12
		Liberia	12700	41.66
		Guinea	1300	4.22
Cestos	15000	Liberia	12800	84.99
		Ivory Coast	2200	14.91
		Guinea	20	0.11
Chiloango	11600	Congo, Democratic Republic of (Kinshasa)	7500	64.6
		Angola	3800	32.71
		Congo, Republic of the (Brazzaville)	300	2.69
Congo/Zaire (4, 5)	3691000	Congo, Democratic Republic of (Kinshasa)	2302800	62.39
		Central African Republic	400800	10.86
		Angola	290600	7.87
		Congo, Republic of the (Brazzaville)	248100	6.72
		Zambia	176000	4.77
		Tanzania, United Republic of	166300	4.51
		Cameroon	85200	2.31
		Burundi	14400	0.39
		Rwanda	4500	0.12
		Sudan	1400	0.04
		Gabon	500	0.01
		Malawi	100	0
		Uganda	70	0
Corubal	24000	Guinea	17500	72.71
		Guinea-Bissau	6500	27.02
Cross	52800	Nigeria	40300	76.34
		Cameroon	12500	23.66
Cuvelai/Etosha	167400	Namibia	114100	68.15
		Angola	53300	31.85
Daoura	34500	Morocco	18200	52.72
		Algeria	16300	47.28
Dra	96400	Morocco	75800	78.65
		Algeria	20600	21.33
Gambia	69900	Senegal	50700	72.48
		Guinea	13200	18.92
		Gambia	5900	8.51
Gash	40000	Eritrea	21400	53.39
		Sudan	9600	24.09
		Ethiopia	9000	22.52
Geba	12800	Guinea-Bissau	8700	67.69
		Senegal	4100	31.88
		Guinea	50	0.42

Continued

Basin name	Total area of basin (sq. km) (1)	Country name	Area of country in basin (sq. km)	Percent area of country in basin (%)
Great Scarcies	12100	Guinea	9000	74.96
		Sierra Leone	3000	25.04
Guir	78900	Algeria	61200	77.53
		Morocco	17700	22.47
Incomati (6)	46700	South Africa	29200	62.47
		Mozambique	14600	31.2
		Swaziland	3000	6.33
Juba-Shibeli	803500	Ethiopia	367400	45.72
		Somalia	220900	27.49
		Kenya	215300	26.79
Komoe	78100	Ivory Coast	58300	74.67
		Burkina Faso	16900	21.66
		Ghana	2200	2.85
		Mali	600	0.82
Kunene	110000	Angola	95300	86.68
		Namibia	14700	13.32
Lake Chad (7)	2388700	Chad	1079200	45.18
		Niger	674200	28.23
		Central African Republic	218600	9.15
		Nigeria	180200	7.54
		Algeria	90000	3.77
		Sudan	82800	3.47
		Cameroon	46800	1.96
		Chad, claimed by Libya	12300	0.51
		Libya	4600	0.19
Lake Natron	55400	Tanzania, United Republic of	37100	67
		Kenya	18300	33
Lake Turkana (8)	206900	Ethiopia	113200	54.69
		Kenya	89700	43.36
		Uganda	2500	1.21
		Sudan	1500	0.7
		Sudan, administered by Kenya	70	0.03
Limpopo	414800	South Africa	183500	44.25
		Mozambique	87200	21.02
		Botswana	81500	19.65
		Zimbabwe	62600	15.08
Little Scarcies	18900	Sierra Leone	13000	69.12
		Guinea	5800	30.88
Loffa	11400	Liberia	10100	88.56
		Guinea	1300	11.38
Lotagipi Swamp (8)	38700	Kenya	20300	52.33
		Sudan	9900	25.54
		Sudan, administered by Kenya	3300	8.52
		Ethiopia	3200	8.32
		Uganda	2100	5.29

Basin name	Total area of basin (sq. km) (1)	Country name	Area of country in basin (sq. km)	Percent area of country in basin (%)
Mana-Morro	6800	Liberia	5700	82.84
		Sierra Leone	1200	17.16
Maputo (6)	30700	South Africa	18500	60.31
		Swaziland	10600	34.71
		Mozambique	1500	4.98
Mbe	7000	Gabon	6500	92.97
		Equatorial Guinea	500	7.02
Medjerda	23100	Tunisia	15600	67.53
		Algeria	7600	32.9
Moa	22500	Sierra Leone	10800	47.79
		Guinea	8800	39.2
		Liberia	2900	13.01
Mono	23400	Togo	22300	95.19
		Benin	1100	4.81
Niger	2113200	Nigeria	561900	26.59
		Mali	540700	25.58
		Niger	497900	23.56
		Algeria	161300	7.63
		Guinea	95900	4.54
		Cameroon	88100	4.17
		Burkina Faso	82900	3.93
		Benin	45300	2.14
		Ivory Coast	22900	1.08
		Chad	16400	0.78
		Sierra Leone	50	0
Nile (9)	3031700	Sudan	1927300	63.57
		Ethiopia	356000	11.74
		Egypt	272600	8.99
		Uganda	238500	7.87
		Tanzania, United Republic of	120200	3.96
		Kenya	50900	1.68
		Congo, Democratic Republic of (Kinshasa)	21400	0.71
		Rwanda	20700	0.68
		Burundi	12900	0.43
		Egypt, administered by Sudan	4400	0.15
		Eritrea	3500	0.12
		Sudan, administered by Egypt	2000	0.07
		Central African Republic	1200	0.04
Nyanga	12300	Gabon	11500	93.56
		Congo, Republic of the (Brazzaville)	800	6.44

Continued

Basin name	Total area of basin (sq. km) (1)	Country name	Area of country in basin (sq. km)	Percent area of country in basin (%)
Ogooue	223000	Gabon	189500	84.98
		Congo, Republic of the (Brazzaville)	26300	11.79
		Cameroon	5200	2.34
		Equatorial Guinea	2000	0.89
Okavango	706900	Botswana	358000	50.65
		Namibia	176200	24.93
		Angola	150100	21.23
		Zimbabwe	22600	3.19
Orange (6, 10, 11)	945500	South Africa	563900	59.65
		Namibia	240200	25.4
		Botswana	121400	12.85
		Lesotho	19900	2.1
Oued Bon Naima	500	Morocco	300	65.08
		Algeria	200	34.92
Oueme	59500	Benin	49400	82.98
		Nigeria	9700	16.29
		Togo	400	0.73
Ruvuma (12)	151700	Mozambique	99000	65.27
		Tanzania, United Republic of	52200	34.43
		Malawi	400	0.3
Sabi	115700	Zimbabwe	85400	73.85
		Mozambique	30300	26.15
Sassandra	68200	Ivory Coast	59800	87.64
		Guinea	8400	12.36
Senegal	436000	Mauritania	219100	50.25
		Mali	150800	34.59
		Senegal	35200	8.08
		Guinea	30800	7.07
St. John (Africa)	15600	Liberia	12900	83.04
		Guinea	2600	16.96
St. Paul	21200	Liberia	11800	55.75
		Guinea	9400	44.25
Tafna	9500	Algeria	7000	74.39
		Morocco	2400	25.6
Tano	15600	Ghana	13700	87.96
		Ivory Coast	1700	11.21
Umba	8200	Tanzania, United Republic of	6800	83.58
		Kenya	1300	16.41
Umbeluzi (6)	10900	Mozambique	7200	65.87
		Swaziland	3500	32.44
		South Africa	30	0.27
Utamboni	7700	Gabon	4500	58.65
		Equatorial Guinea	3100	40.4

Basin name	Total area of basin (sq. km) (1)	Country name	Area of country in basin (sq. km)	Percent area of country in basin (%)
Volta	412800	Burkina Faso	173500	42.04
		Ghana	166000	40.21
		Togo	25800	6.26
		Mali	18800	4.56
		Benin	15000	3.63
		Ivory Coast	13500	3.27
Zambezi (13, 14)	1385300	Zambia	576900	41.64
		Angola	254600	18.38
		Zimbabwe	215500	15.55
		Mozambique	163500	11.81
		Malawi	110400	7.97
		Tanzania, United Republic of	27200	1.97
		Botswana	18900	1.37
		Namibia	17200	1.24
		Congo, Democratic Republic of (Kinshasa)	1100	0.08

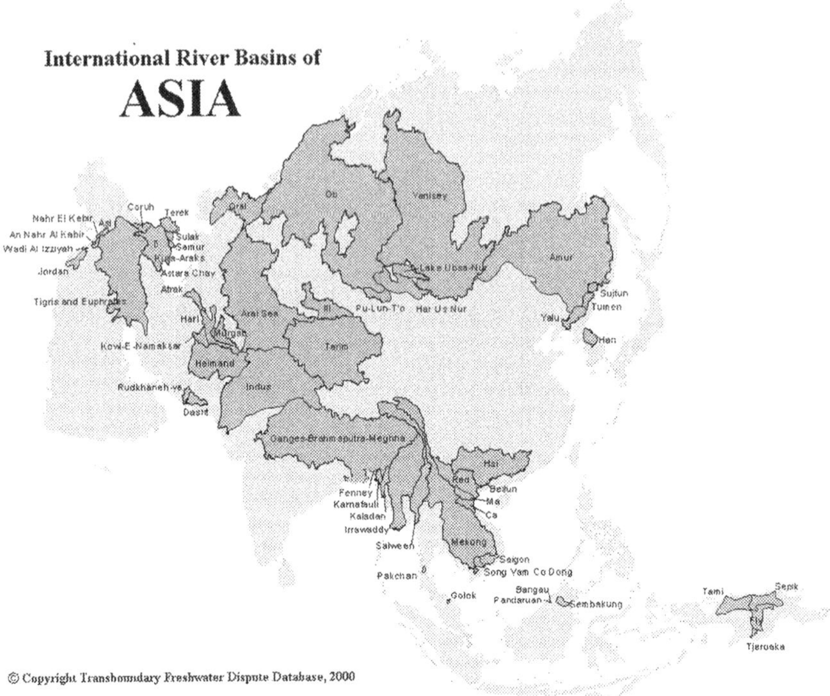

Basin name	Total area of basin (sq. km) (1)	Country name	Area of country in basin (sq. km)	Percent area of country in basin (%)
Amur (15)	2085900	Russia	1006100	48.23
		China	889100	42.62
		Mongolia	190600	9.14
		Korea, Democratic People's Republic of (North)	100	0.01
An Nahr Al Kabir	1300	Syria	900	67.6
		Lebanon	400	31.7
Aral Sea (16, 17)	1231400	Kazakhstan	424400	34.46
		Uzbekistan	382600	31.07
		Tajikistan	135700	11.02
		Kyrgyzstan	111700	9.07
		Afghanistan	104900	8.52
		Turkmenistan	70000	5.68
		China	1900	0.15
		Pakistan	200	0.01
Asi/Orontes	37900	Turkey	18900	49.94
		Syria	16800	44.32
		Lebanon	2200	5.74
Astara Chay (18)	600	Iran	500	81.64
		Azerbaijan	100	18.36

Basin name	Total area of basin (sq. km) (1)	Country name	Area of country in basin (sq. km)	Percent area of country in basin (%)
Atrak (17)	34200	Iran	23600	68.86
		Turkmenistan	10700	31.14
BahuKalat/ Rudkhanehye	18000	Iran	18000	99.83
		Pakistan	30	0.17
Bangau (19)	60	Brunei	30	46.03
		Malaysia	30	49.21
Bei Jiang/His (20)	417800	China	407900	97.63
		Vietnam	9800	2.35
Beilun (20)	900	China	800	84.92
		Vietnam	100	15.08
Ca/Song Koi	31000	Vietnam	20100	64.91
		Laos, People's Democratic Republic of	10900	35.09
Coruh (18)	22100	Turkey	20000	90.85
		Georgia	2000	9.01
Dasht	33400	Pakistan	26200	78.42
		Iran	7200	21.58
Fenney	2800	India	1800	65.83
		Bangladesh	1000	34.17
Fly	64600	Papua New Guinea	60400	93.4
		Indonesia	4300	6.6
Ganges-Brahmaputra-Meghna (21, 22)	1634900	India	948400	58.01
		China	321300	19.65
		Nepal	147400	9.01
		Bangladesh	107100	6.55
		India, claimed by China	67100	4.11
		Bhutan	39900	2.44
		India control, claimed by China	1200	0.07
		Myanmar (Burma)	80	0
Golok	1800	Thailand	1000	56.62
		Malaysia	800	43.38
Han (23, 24)	35300	Korea, Republic of (South)	25100	71.22
		Korea, Democratic People's Republic of (North)	10100	28.67
Har Us Nur	185300	Mongolia	179300	96.81
		Russia	5600	3.04
		China	300	0.15
Hari/Harirud	92600	Afghanistan	41000	44.31
		Iran	35400	38.27
		Turkmenistan	16100	17.41
Helmand	353500	Afghanistan	288200	81.53
		Iran	54900	15.52
		Pakistan	10400	2.95

Continued

Basin name	Total area of basin (sq. km) (1)	Country name	Area of country in basin (sq. km)	Percent area of country in basin (%)
Ili/Kunes He	161200	Kazakhstan	97100	60.24
		China	55300	34.32
		Kyrgyzstan	8800	5.44
Indus (25, 26)	1138800	Pakistan	597700	52.48
		India	381600	33.51
		China	76200	6.69
		Afghanistan	72100	6.33
		Chinese control, claimed by India	9600	0.84
		Indian control, claimed by China	1600	0.14
		Nepal	10	0
Irrawaddy	404200	Myanmar (Burma)	368600	91.2
		China	18500	4.58
		India	14100	3.49
		India, claimed by China	1200	0.3
Jenisej/Yenisey	2557800	Russia	2229800	87.17
		Mongolia	327900	12.82
Jordan (27, 28, 29)	42800	Jordan	20600	48.13
		Israel	9100	21.26
		Syria	4900	11.45
		West Bank	3200	7.48
		Egypt	2700	6.31
		Golan Heights	1500	3.5
		Lebanon	600	1.33
Kaladan	30500	Myanmar (Burma)	22900	74.91
		India	7300	23.84
Karnaphuli	12500	Bangladesh	7400	58.78
		India	5100	41.14
		Myanmar (Burma)	10	0.09
Kowl E Namaksar	36500	Iran	25900	71.13
		Afghanistan	10500	28.87
Kura-Araks (18)	193200	Azerbaijan	56600	29.28
		Iran	39700	20.55
		Armenia	34800	18.03
		Georgia	34300	17.77
		Turkey	27700	14.32
		Russia	60	0.03
Lake Ubsa-Nur	62800	Mongolia	47600	75.78
		Russia	15200	24.22
Ma	30300	Vietnam	17100	56.48
		Laos, People's Democratic Republic of	13200	43.52
Mekong (30, 31)	787800	Laos, People's Democratic Republic of	198000	25.14
		Thailand	193900	24.62
		China	171700	21.79

Basin name	Total area of basin (sq. km) (1)	Country name	Area of country in basin (sq. km)	Percent area of country in basin (%)
		Cambodia (Kampuchea)	158400	20.1
		Vietnam	38200	4.84
		Myanmar (Burma)	27600	3.51
Murgab	60900	Afghanistan	36400	59.79
		Turkmenistan	24500	40.21
Nahr El Kebir	1500	Syria	1300	85.61
		Turkey	200	13.87
Ob (18)	2950800	Russia	2192700	74.31
		Kazakhstan	743800	25.21
		China	13900	0.47
		Mongolia	200	0.01
Oral/Ural (18)	311000	Kazakhstan	175500	56.43
		Russia	135500	43.57
Pakchan	3900	Myanmar (Burma)	1900	49.11
		Thailand	1800	47.24
Pandaruan (19)	400	Brunei	200	60.65
		Malaysia	100	39.08
Pu Lun T'o	89000	China	77800	87.39
		Mongolia	11100	12.48
		Russia	80	0.09
		Kazakhstan	30	0.04
Red/Song Hong (20)	157100	China	84500	53.75
		Vietnam	71500	45.5
		Laos, People's Democratic Republic of	1200	0.74
Saigon	25100	Vietnam	24800	98.67
		Cambodia (Kampuchea)	200	0.99
Salween	244000	China	127900	52.4
		Myanmar (Burma)	107000	43.85
		Thailand	9100	3.73
Samur (18)	6800	Russia	6300	93.75
		Azerbaijan	400	6.22
Sembakung (19)	15300	Indonesia	8100	52.86
		Malaysia	7200	47.14
Sepik	73400	Papua New Guinea	71000	96.81
		Indonesia	2300	3.19
Song Vam Co Dong	15300	Vietnam	7800	50.68
		Cambodia (Kampuchea)	7500	49.23
Sujfun	18300	China	11800	64.46
		Russia	6500	35.54
Sulak (18)	15100	Russia	13900	92.38
		Georgia	1100	7.24
		Azerbaijan	60	0.38

Continued

Basin name	Total area of basin (sq. km) (1)	Country name	Area of country in basin (sq. km)	Percent area of country in basin (%)
Tami	89900	Indonesia	87700	97.55
		Papua New Guinea	2200	2.45
Tarim (16, 17, 25, 26)	1051600	China	1000300	95.12
		Chinese control, claimed by India	21500	2.04
		Kyrgyzstan	21100	2
		Tajikistan	6600	0.63
		Pakistan	2000	0.19
		Afghanistan	60	0.01
Terek (18)	38700	Russia	37000	95.39
		Georgia	1800	4.61
Tigris-Euphrates/Shatt al Arab (32)	789000	Iraq	319400	40.48
		Turkey	195700	24.8
		Iran	155400	19.7
		Syria	116300	14.73
		Jordan	2000	0.25
		Saudi Arabia	80	0.01
Tjeroaka-Wanggoe	6600	Indonesia	4000	61.57
		Papua New Guinea	2500	38.43
Tumen	29100	China	20300	69.75
		Korea, Democratic People's Republic of (North)	8300	28.59
		Russia	500	1.66
Wadi Al Izziyah	600	Lebanon	400	68.23
		Israel	200	31.6
Yalu	50900	China	26800	52.65
		Korea, Democratic People's Republic of (North)	23800	46.82

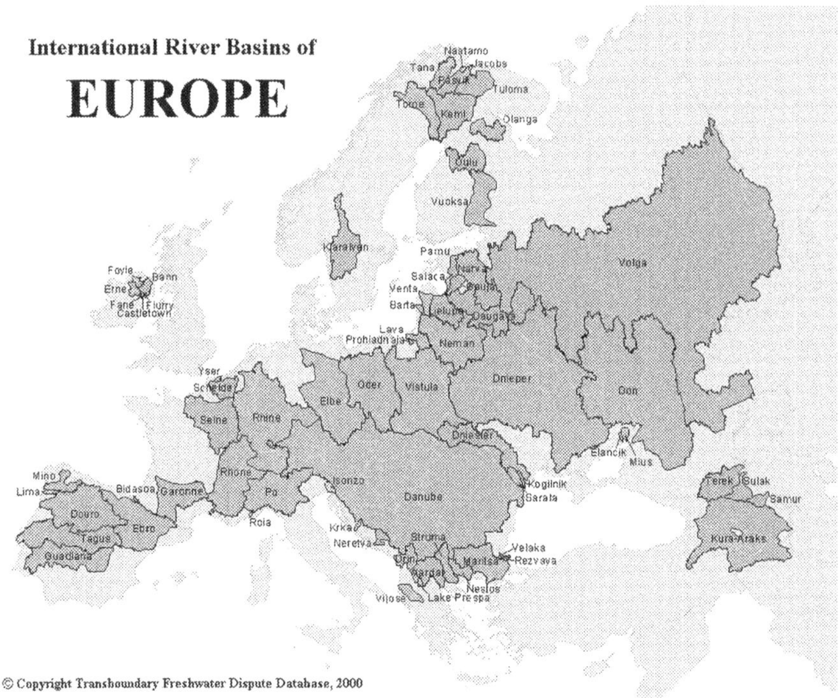

Basin name	Total area of basin (sq. km) (1)	Country name	Area of country in basin (sq. km)	Percent area of country in basin (%)
Bann	5600	United Kingdom	5400	97.14
		Ireland	200	2.86
Barta	1800	Latvia	1100	60.87
		Lithuania	700	37.71
Bidasoa	500	Spain	500	89.33
		France	60	10.67
Castletown	400	United Kingdom	300	76.12
		Ireland	90	23.88
Danube (33, 34, 35, 36, 37)	790100	Romania	228500	28.93
		Hungary	92800	11.74
		Austria	81600	10.32
		Yugoslavia (Serbia and Montenegro)	81500	10.31
		Germany	59000	7.47
		Slovakia	45600	5.77
		Bulgaria	40900	5.17
		Bosnia and Herzegovina	38200	4.83
		Croatia	35900	4.54

Continued

Basin name	Total area of basin (sq. km) (1)	Country name	Area of country in basin (sq. km)	Percent area of country in basin (%)
		Ukraine	29600	3.75
		Czech Republic	20500	2.59
		Slovenia	17200	2.18
		Moldova	13900	1.76
		Switzerland	2500	0.32
		Italy	1200	0.15
		Poland	700	0.09
		Albania	200	0.03
Daugava (38, 39)	58700	Byelarus	28300	48.14
		Latvia	20200	34.38
		Russia	9500	16.11
		Lithuania	800	1.38
Dnieper	516300	Ukraine	299300	57.97
		Byelarus	124900	24.19
		Russia	92100	17.83
Dniester (37)	62000	Ukraine	46800	75.44
		Moldova	15200	24.52
		Poland	30	0.05
Don	425600	Russia	371200	87.23
		Ukraine	54300	12.76
Douro/Duero	98900	Spain	80700	81.63
		Portugal	18200	18.37
Drin (36)	17900	Albania	8100	45.39
		Yugoslavia (Serbia and Montenegro)	7400	41.4
		Macedonia	2200	12.18
Ebro	85800	Spain	85200	99.36
		Andorra	400	0.48
		France	100	0.16
Elancik	900	Russia	700	71.32
		Ukraine	300	28.68
Elbe	132200	Germany	83100	62.86
		Czech Republic	47600	36.02
		Austria	700	0.54
		Poland	700	0.56
Erne	4800	Ireland	2800	59.28
		United Kingdom	1900	40.72
Fane	200	Ireland	200	96.46
		United Kingdom	10	3.54
Flurry	60	United Kingdom	50	73.77
		Ireland	20	26.23
Foyle	2900	United Kingdom	2000	67.3
		Ireland	1000	32.7
Garonne	55800	France	55100	98.83
		Spain	600	1.07
		Andorra	40	0.08
Gauja	11600	Latvia	10400	90.42
		Estonia	1100	9.58

Basin name	Total area of basin (sq. km) (1)	Country name	Area of country in basin (sq. km)	Percent area of country in basin (%)
Glama	43000	Norway	42600	99
		Sweden	400	0.99
Guadiana	67900	Spain	54900	80.82
		Portugal	13000	19.18
Isonzo	3000	Slovenia	1800	59.48
		Italy	1200	40.09
Jacobs	400	Norway	300	68.1
		Russia	100	31.9
Kemi	55700	Finland	52700	94.52
		Russia	3000	5.41
		Norway	10	0.01
Klaralven	51000	Sweden	43100	84.54
		Norway	7900	15.46
Kogilnik (37)	6100	Moldova	3600	57.82
		Ukraine	2600	42.18
Krka	1300	Croatia	1100	89.55
		Bosnia and Herzegovina	100	8.93
		Yugoslavia (Serbia and Montenegro)	10	0.4
Lake Prespa	9000	Albania	8000	88.17
		Macedonia	800	8.5
		Greece	300	3.32
Lava/Pregel	8600	Russia	6300	74
		Poland	2000	23.84
Lielupe	14400	Latvia	9600	66.76
		Lithuania	4800	33.22
Lima	2300	Spain	1200	50.88
		Portugal	1100	49.04
Maritsa	49600	Bulgaria	33000	66.49
		Turkey	12800	25.69
		Greece	3700	7.55
Mino	15100	Spain	14500	96.18
		Portugal	600	3.7
Mius	2800	Russia	1900	69.82
		Ukraine	800	30.07
Naatamo	1000	Norway	600	57.73
		Finland	400	41.97
Narva (40, 41)	53000	Russia	28200	53.2
		Estonia	18100	34.09
		Latvia	5900	11.13
		Byelarus	800	1.57
Neman (38, 39)	90300	Byelarus	41700	46.13
		Lithuania	39700	43.97
		Russia	4800	5.3
		Poland	3800	4.21
		Latvia	300	0.36

Continued

Basin name	Total area of basin (sq. km) (1)	Country name	Area of country in basin (sq. km)	Percent area of country in basin (%)
Neretva	5500	Bosnia and Herzegovina	5300	95.98
		Croatia	200	3.47
Nestos	10200	Bulgaria	5500	53.63
		Greece	4700	46.36
Oder/Odra	122400	Poland	103100	84.2
		Czech Republic	10300	8.38
		Germany	7800	6.33
		Slovakia	1300	1.09
Olanga	18800	Russia	16800	89.37
		Finland	2000	10.62
Oulu	28700	Finland	26700	93.2
		Russia	1900	6.78
Parnu	5800	Estonia	5800	99.85
		Latvia	10	0.15
Pasvik	16000	Finland	12400	77.46
		Russia	2600	16.15
		Norway	1000	6.39
Po	87100	Italy	82200	94.44
		Switzerland	4300	4.92
		France	500	0.54
		Austria	90	0.1
Prohladnaja	600	Russia	500	76.9
		Poland	100	23.1
Rezvaya	700	Turkey	500	74.66
		Bulgaria	200	25.34
Rhine (42)	172900	Germany	97700	56.49
		Switzerland	24300	14.05
		France	23100	13.34
		Belgium	13900	8.03
		Netherlands	9900	5.75
		Luxembourg	2500	1.46
		Austria	1300	0.76
		Liechtenstein	200	0.09
		Italy	70	0.04
Rhone	100200	France	90100	89.88
		Switzerland	10100	10.05
		Italy	50	0.05
Roia	600	France	400	67.39
		Italy	200	30.45
Salaca	2100	Latvia	1600	78.52
		Estonia	100	5.7
Sarata (37)	1800	Ukraine	1100	63.9
		Moldova	600	36.05
Schelde	17100	France	8600	50.03
		Belgium	8400	49.28
		Netherlands	80	0.47

Basin name	Total area of basin (sq. km) (1)	Country name	Area of country in basin (sq. km)	Percent area of country in basin (%)
Seine	85700	France	83800	97.78
		Belgium	1800	2.14
		Luxembourg	70	0.08
Struma (36)	15000	Bulgaria	8600	57.66
		Greece	3900	25.88
		Macedonia	1800	12.22
		Yugoslavia (Serbia and Montenegro)	600	4.19
Tagus/Tejo	77900	Spain	51400	66.06
		Portugal	26100	33.5
Tana	15600	Norway	9300	59.71
		Finland	6300	40.23
Torne/Tornealven	37300	Sweden	25400	67.98
		Finland	10400	28
		Norway	1500	4.03
Tuloma	25800	Russia	23700	91.85
		Finland	2000	7.93
Vardar (36)	32400	Macedonia	20300	62.83
		Yugoslavia (Serbia and Montenegro)	8200	25.22
		Greece	3900	11.94
Velaka	700	Bulgaria	700	95.25
		Turkey	30	3.74
Venta	9500	Latvia	6200	65.15
		Lithuania	3300	34.72
Vijose	7200	Albania	4600	64.83
		Greece	2500	34.66
Vistula/Wista	194000	Poland	169700	87.45
		Ukraine	12700	6.55
		Byelarus	9800	5.03
		Slovakia	1900	0.96
		Czech Republic	20	0.01
Volga (18)	1554900	Russia	1551300	99.77
		Kazakhstan	2200	0.14
		Byelarus	1300	0.08
Vuoksa	62700	Finland	54300	86.48
		Russia	8500	13.52
Wiedau	1100	Denmark	1000	86.23
		Germany	200	13.32
Yser	900	France	500	53.63
		Belgium	400	46.37

International River Basins of

NORTH AMERICA

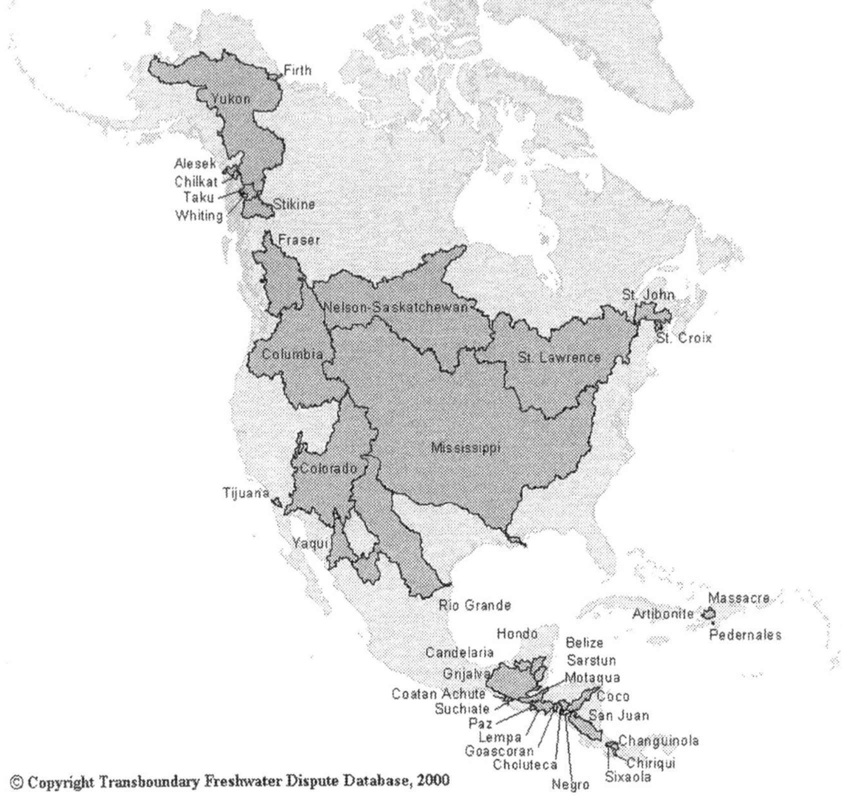

Basin name	Total area of basin (sq. km) (1)	Country name	Area of country in basin (sq. km)	Percent area of country in basin (%)
Alsek	28400	Canada	26500	93.5
		United States of America of America	1800	6.5
Artibonite	8800	Haiti	6600	74.37
		Dominican Republic	2300	25.55
Belize (43)	11500	Belize	7000	60.86
		Guatemala	4500	39.14
Candelaria	12800	Mexico	11300	88.24
		Guatemala	1500	11.74
Changuinola	3200	Panama	2900	91.29
		Costa Rica	300	8.33
Chilkat	3800	United States of America	2100	56.59
		Canada	1600	43.35

Basin name	Total area of basin (sq. km) (1)	Country name	Area of country in basin (sq. km)	Percent area of country in basin (%)
Chiriqui	1700	Panama	1500	86.17
		Costa Rica	200	13.83
Choluteca	7400	Honduras	7200	97.68
		Nicaragua	200	2.32
Coatan Achute	2000	Mexico	1700	86.27
		Guatemala	300	13.73
Coco/Segovia	25400	Nicaragua	17900	70.52
		Honduras	7500	29.48
Colorado	655000	United States of America	644600	98.41
		Mexico	10400	1.59
Columbia	668400	United States of America	566500	84.75
		Canada	101900	15.24
Firth	6000	Canada	3800	63.6
		United States of America	2200	36.4
Fraser	239700	Canada	239100	99.74
		United States of America	600	0.26
Goascoran	2800	Honduras	1500	53.36
		El Salvador	1300	46.64
Grijalva (43)	126800	Mexico	78900	62.25
		Guatemala	47800	37.72
		Belize	20	0.02
Hondo (43)	14600	Mexico	8900	61.14
		Guatemala	4200	28.5
		Belize	1500	10.36
Lempa	18000	El Salvador	9500	52.45
		Honduras	5800	32.01
		Guatemala	2800	15.54
Massacre	800	Haiti	500	62.03
		Dominican Republic	300	35.96
Mississippi	3226300	United States of America	3176500	98.46
		Canada	49800	1.54
Motaqua	16100	Guatemala	14600	90.85
		Honduras	1500	9.11
Negro	5800	Nicaragua	4800	83.87
		Honduras	900	15.68
Nelson-Saskatchewan	1109400	Canada	952000	85.81
		United States of America	157400	14.19
Paz	2200	Guatemala	1400	64.47
		El Salvador	800	35.53
Pedernales	400	Haiti	200	67.32
		Dominican Republic	100	32.68

Continued

Basin name	Total area of basin (sq. km) (1)	Country name	Area of country in basin (sq. km)	Percent area of country in basin (%)
Rio Grande (North America)	656100	United States of America	341800	52.1
		Mexico	314300	47.9
San Juan	42200	Nicaragua	30400	72.02
		Costa Rica	11800	27.93
Sarstun (43)	2100	Guatemala	1800	87.63
		Belize	300	12.37
Sixaola	2900	Costa Rica	2500	88.65
		Panama	300	9.99
Skagit	8000	United States of America	7100	88.46
		Canada	900	11.54
St. Croix	4600	United States of America	3300	70.86
		Canada	1400	29.14
St. John (North America)	47700	Canada	30300	63.5
		United States of America	17300	36.22
St. Lawrence	1055200	Canada	559000	52.98
		United States of America	496100	47.02
Stikine	50900	Canada	50000	98.32
		United States of America	900	1.68
Suchiate	1600	Guatemala	1100	68.79
		Mexico	500	31.21
Taku	18100	Canada	16300	90.09
		United States of America	1700	9.13
Tijuana	4400	Mexico	3100	70.57
		United States of America	1300	29.43
Whiting	2600	Canada	2000	80.06
		United States of America	500	19.94
Yaqui	74700	Mexico	70100	93.87
		United States of America	4600	6.13
Yukon	829700	United States of America	496400	59.83
		Canada	333300	40.17

International River Basins of

SOUTH AMERICA

Basin name	Total area of basin (sq. km) (1)	Country name	Area of country in basin (sq. km)	Percent area of country in basin (%)
Amacuro	5600	Venezuela	4900	86.89
		Guyana	700	13.11
Amazon (44)	5883400	Brazil	3670300	62.38
		Peru	956500	16.26
		Bolivia	706700	12.01
		Colombia	367800	6.25
		Ecuador	123800	2.1
		Venezuela	40300	0.68
		Guyana	14500	0.25
		Suriname	1400	0.02
		French Guiana	30	0
Aviles	300	Argentina	200	88.72
		Chile	30	11.28
Aysen	13600	Chile	13100	96.07
		Argentina	500	3.93
Baker	30800	Chile	21000	68.15
		Argentina	9800	31.83
Barima	2100	Guyana	1100	51.05
		Venezuela	1000	47.84
Cancoso/Lauca	23500	Bolivia	20200	85.72
		Chile	3400	14.28
Carmen Silva/Chico	1700	Argentina	1000	59.7
		Chile	700	40.3
Catatumbo	31000	Colombia	19600	63.15
		Venezuela	11400	36.75
Chira (44)	15700	Peru	9800	62.23
		Ecuador	5800	37.23
Chuy	200	Brazil	100	64.57
		Uruguay	60	32.57
Comau	900	Chile	900	91.36
		Argentina	80	8.64
Corantijn/ Courantyne (45)	41800	Guyana	21700	52.06
		Suriname	19900	47.75
		Brazil	80	0.19
Cullen	600	Chile	500	83
		Argentina	100	17
Essequibo (46)	239500	Guyana	162100	67.67
		Venezuela	52400	21.87
		Suriname	24300	10.13
		Brazil	200	0.07
Gallegos-Chico	11600	Argentina	7000	60.15
		Chile	4600	39.85
Jurado	700	Colombia	500	82.11
		Panama	100	17.89
La Plata (47, 48)	2954500	Brazil	1379300	46.69
		Argentina	817900	27.68
		Paraguay	400100	13.54
		Bolivia	245100	8.3
		Uruguay	111600	3.78

Basin name	Total area of basin (sq. km) (1)	Country name	Area of country in basin (sq. km)	Percent area of country in basin (%)
Lagoon Mirim	55000	Uruguay	31200	56.69
		Brazil	23800	43.24
Lake Fagnano (49)	3200	Argentina	2700	85.17
		Chile	500	14.83
Lake Titicaca-Poopo System	111800	Bolivia	63000	56.32
		Peru	48000	42.94
		Chile	800	0.74
Maroni (50)	65000	Suriname	37500	57.64
		French Guiana	27200	41.9
		Brazil	200	0.27
Mataje	700	Ecuador	500	73.98
		Colombia	200	26.02
Mira	12100	Colombia	6200	50.87
		Ecuador	5800	47.97
Oiapoque/ Oyupock	23300	French Guiana	13700	58.92
		Brazil	9500	41
Orinoco	927400	Venezuela	604500	65.18
		Colombia	321700	34.68
		Brazil	800	0.08
Palena	13300	Chile	7300	54.87
		Argentina	6000	45.13
Pascua	13700	Chile	7300	53.51
		Argentina	6400	46.46
Patia	21300	Colombia	20800	97.61
		Ecuador	500	2.38
Puelo	8400	Argentina	5500	66.03
		Chile	2900	33.97
Rio Grande (South America)	8000	Argentina	4000	49.74
		Chile	4000	50.26
San Martin	700	Chile	600	87.44
		Argentina	80	12.56
Seno Union/Serrano	6500	Chile	5700	87.93
		Argentina	700	10.34
Tumbes-Poyango (44)	5000	Ecuador	3600	71.62
		Peru	1400	28.38
Valdivia	15000	Chile	14700	98.39
		Argentina	100	0.69
Yelcho	11100	Argentina	6900	62.14
		Chile	4200	37.86
Zapaleri (51)	2600	Chile	1600	59.6
		Argentina	500	19.65
		Bolivia	500	20.75
Zarumilla (44)	4300	Ecuador	3400	78.71
		Peru	900	20.51

Index